2,637 YEARS OF
PHYSICS
FROM THALES OF MILETOS TO THE MODERN ERA

SHELDON COHEN

Copyright 2017 Sheldon Cohen,
All rights reserved.

Published by eBookIt.com

ISBN-13: 978-1-4566-2915-1

No part of this book may be reproduced in any form or by any electronic or mechanical means including information storage and retrieval systems, without permission in writing from the author. The only exception is by a reviewer, who may quote short excerpts in a review.

The author, Sheldon Cohen,
is a retired physician with a lifetime
fascination of Physics.

So, why not write a book?

Dedicated to
BETTY,
Gail, Paul,
Amanda, Shane, Megan, Travis, Carly, Alexa,
Ethan, Emily,
Derek, Rylie, Benjamin.
ILYI

CONTENTS

THE FIRST THOUGHT EXPERIMENT: IS THERE A PRIMORDIAL BUILDING BLOCK?..1

WHAT IS LIGHT?..12

THERE ARE ATOMS AFTER ALL 14

ATOMS ELEMENTS COMPOUNDS.................20

WHAT IS THIS ATTRACTIVE FORCE? 23

BEAUTIFUL LIGHTS. FORCES OF ATTRACTION AND REPULSION. MYSTERIOUS RAYS..32

THE FIRST LOOK INSIDE THE ATOM…X RAYS..54

A FURTHER LOOK INSIDE THE ATOM… RADIOACTIVITY 56

BACK TO LIGHT 61

THE UNIVERSE IN MOTION 63

THE SPECTRUM: AN INDIRECT PEEK INSIDE THE ATOM 73

ANOTHER PEEK INSIDE THE ATOM---THE ELECTRON---WHAT DOES AN ATOM LOOK LIKE? 77

A BETTER VIEW OF THE ATOM 80

OPEN THE DOOR TO A NEW PHYSICS 84

A NEW ATOM. THE START OF QUANTUM MECHANICS 99

A NEW GRAVITY 106

BACK TO THE ATOM 113

PARTICLE AND WAVE 119

IT'S FUZZY DOWN THERE 122

THE NUCLEAR AGE: PROMISE OR DISASTER? 131

END GAME 136

WHAT HAPPENED TO AIR, EARTH, FIRE AND WATER? 146

    HOW THE UNIVERSE WORKS 146

    SUBATOMIC PARTICLES 146

HISTORY OF THE UNIVERSE 150

THE FIRST THOUGHT EXPERIMENT: IS THERE A PRIMORDIAL BUILDING BLOCK?

Once humankind developed the capability to contemplate who they were in relation to the world about them, some began to wonder if there was a similarity to all objects in the environment. It was easy to notice that the differences were visually vast, but could there be a common indivisible building block for all the great variety of animate and inanimate objects that were visible to our senses?

For instance, some of these thinkers wondered: if you were able to cut a gold bar in half, and then cut the half piece in half, and then cut the quarter piece in half, and then the eighth, and then the sixteenth, etcetera, when are you forced to stop; or will you ever reach that point? Is there a time in this process when you would end up with a fragment that could no longer be broken down into a smaller piece (the smallest possible piece still identifiable as gold)? The Greeks named this infinitesimally tiny end piece *atomos* (atom) meaning 'unbreakable' in the Greek language.

In this era, proof of the makeup of the microscopic world did not exist in the minds of man, so the level of the infinitesimally small could only be conjectured upon via mythological thought experiments. Thus it would remain for many centuries until the invention of the microscope opened up this realm to human view and slowly expanded thought.

Although the atomistic concept, devised strictly through the minds of human beings, was a brilliant thought experiment and portended great advances to come, there were still those who disputed this theory and felt that the animate and inanimate objects of the world were visually accessible and made of materials that were common to our senses. We could see these primary substances. Everything in our sight was visually accessible, and formed by combining together in a variety of amounts that made for almost infinite combinations. It would be this mythological thought that would remain supreme in the minds of man for the centuries it would take before a return to the atomistic concept and the rapid advances which would finally ensue in the minds of advanced thinkers. Who were these great pioneers of ancient times?

Thales of Miletos (624-546 BCE)
He is one of the first to attempt an evaluation about the above mentioned theory. He is recognized as the founder of ancient Greek philosophy and physics, Thales suggested that water was the most fundamental structure that led to all other things. His reasoning: the ability of water to become vapor and exhibit motion. According to the Greeks, this capacity for change and motion indicated life. Thales felt that our universe was a living organism directed by water as the primary substance.

As mentioned before Thales, explanations for any natural phenomena had always relied on mythology, an aspect of religion. Thales contribution to the advancement of human intellect was a first step in the evolution of thinking from myth to reason.

Thales became fascinated by the force of attraction between bodies and made a serious study of this phenomenon. Even primitive man undoubtedly recognized this attractive force, but no one ever studied the phenomenon with the intent of understanding what such attraction meant until Thales.

Miletos, on the Aegean Sea in an area that is now Turkey, was near a town known as Magnesia. Magnesia had abundant deposits of lodestone, a naturally occurring type of iron ore well known to attract iron but no other substance. Someone named this attractive force magnetism and a substance that had this property became a "magnet." Thales appreciated this phenomenon and became fascinated, thus embarking upon a serious study in an attempt to understand what it meant.

As part of this study, Thales discovered that rubbing amber with cat fur caused the amber to attract light objects like feathers or straw in a manner similar to the iron ore. Clearly, this represented a mysterious attractive force, and the Greeks therefore believed that the amber had a 'soul.' They considered this force different from magnetism that only involved iron, because many substances also had the ability to attract other objects when rubbed. Why? The answer would be a long time coming as it would remain a mystery for 2500 years. This would prove to be a major scientific discovery from which great advances evolved; for this attractive force would, centuries later, come to be known as static electricity. Ancient man was good at discerning, but did not have the basis to explain what it was exactly that they had discerned.

Thales was forever steeped in study. Plato tells the story about Thales once falling into a well while focusing his attention on the stars above. A servant girl, who pulled him out, stated that he was so eager to know the stars that he paid no attention to what was under his feet. They wrote on his tomb: *Here in a narrow tomb great Thales lies; yet his renown for wisdom reached the skies.*

We will jump to three centuries later when another Greek philosopher,

Theophrastus (371-286 BCE)

determined that other substances when rubbed shared amber's unusual attractive power. Could this represent a universal force shared by matter? The thinkers of the day wondered what this force was, but at that time they could not advance beyond what was only an interesting discussion.

Back to the fifth and sixth centuries BCE

Anaximander (610-546 BCE)

was one of Thales pupils who disputed Thale's thesis. For his time, he realistically proposed his own thesis: he denied that the fundamental building block was anything of form and structure such as the water of Thale's hypothesis, but rather permeated the world and was infinite and timeless and transformed itself into various forms of matter discernable by our senses. This suggested that the real unifying force was more mysterious than could be visualized and lay currently hidden and undiscovered. This was an intellectual way of saying, "we don't know enough yet."

Aniximander's contribution was to suggest that theories based upon the reality of the world around us, and on our visual senses, were perhaps too primitive in terms of the understanding of the time to suggest any real unifying mechanism. More discoveries were necessary before a solid foundation evolved into factual theories. In the meantime, his suggestion of an infinite and timeless primordial building block was closer to the mark, but again would have a many centuries wait for clarification.

Anaximenes (570-528 BCE)

disputed both Thales and Anaximander, and suggested that air was the primary building block from which all else was created. Here we see another educated, for the times, wild guess. Anaximenes reasoned that since air was continuously in motion, it was life. In a state of even dispersal it presents itself as the air we breathe. In a condensed state it presents as mist, then as water, and finally as solid matter, the density of which depends upon the degree of condensation. Air, he felt, was one aspect of a series of changes from fire to air to wind to cloud to water to earth to stones. He attempted to confirm his thesis by simple experimental observation such as blowing on his hand with lips open widely and with pursed lips. In the first instance the air is warm. In the second instance the air is cold. This suggested that air is warm when rarified and cold when concentrated.

Anaximenes' thesis that natural processes are responsible for the formation and change taking place in our world was an important development in the evolution of scientific thought. His contribution was to be the first to suggest a theory and try to prove it by thoughtful observation. At least this was a step in the right direction. Anaximenes had expanded this centuries old thesis. Others would soon enter the fray, such as....

Heraclitus of Ephesus (533-475 BCE)
added other elements to the century long discussion by insisting that fire was the source of everything. His main thought was to suggest that it also had life as manifested by its ever-changing ability. It evolved from fuel to fire to smoke to clouds to rain to oceans to earth. Heraclitus stood for the thesis that there was a unity of the world, but it depended upon and was consistent with constant change of opposites such as heat and cold, day and night, and life and death. This change suggested equilibrium, and his contribution was to suggest that such equilibrium indicated orderliness in our world.

Heraclitus was considered the most influential Greek philosopher before Socrates. What this deep thinking did to his mind is open to conjecture, however, for he was said to have retreated into the forest where he lived on plants, and tried to cure the dropsy (leg swelling) which he developed by covering himself with manure. It didn't work.

The problem for the ancient Greek philosophers was the difficulty in attempting to develop a theory of a single entity evolving into the great variety of objects in the world.

To reconcile this conundrum,

Empodocles of Sicily (fifth century BCE)
suggested that rather than one basic element there were four: air, fire, earth, and water. These basic four elements, mixed and partially combined and separated, thus resulting in the various familiar forms of matter.

In the future, philosophers who adopted the primary building block theory suggested that the four basic elements of air, fire, earth, and water were each made of different atoms. This somewhat brought the atomos concept back into thought. Even with only four basic building blocks, the various combinations of these could result in a great variety of different forms of matter. If the proportions of the building blocks remained the same, there were twenty-four different combinations. Now consider the infinite number of combinations that would occur if you varied the amounts of each of the four basic elements.

For the first time, a combination of actual basic substances could unite to explain the great variety of forms and events which make up our experience.

In addition to the four basic elements of our earthly experience, there was a fifth: the aether that permeated the heavens as far as the eye could see. The aether, considered an invisible, elastic medium distributed through all space beyond the earth's atmosphere remained a building block of physics for twenty-four centuries. More later.

Anaxagoras (500-428 BCE)
suggested that rather than the four basic elements of earth, air, fire, and water, there were infinite seeds composing all matter and not just the basic four. The proportion of the various seeds explained the great diversity of everything around us. From this concept of infinite seeds, or building blocks, it was a simple leap to the idea that there was a smallest, ultimate basic building block of all matter---the atom. It was postulated that in solid bodies, the atoms were held together by mysterious forces, while in gases, the atoms were separate and free to move in space. Over time, the atomists began to forge ahead as the theory of the four basic substances began to wane.

It was the Greek philosopher…

Leucippus (490-? BCE)
who strongly supported the concept of unbreakable tiny fragments, further promoting the atomos concept.

If traveling to Athens from a northwesterly direction, one will pass the Democritus Nuclear Research Laboratory. Naming this facility for...

Democritus (460-370 BCE)

honors a man whose atomos philosophy comes close to modern physics theory. Although he was not the first to espouse atomism, his use of this concept allowed him to develop a much more detailed, and what would prove to be a much more insightful view of the way the world functioned physically.

He believed that space was a vacuum, but in spite of this property of emptiness it could be thought of as existing as did the visual realities of our world. In the void of space, and in the world around us, there were an infinite number of atoms, so small that they were incapable of further division. These atoms made up the physical world. He postulated that all changes occurring in the universe were merely dependent on the density of the atoms and their movement in relation to each other. Nature itself was nothing more than a complex interaction of atoms that followed the laws of mathematics. Initially atoms moved incoherently, but over time they would randomly interact and combine in a multitude of ways responsible for the origin of the universe and the laws of mechanics and motion. As time would subsequently prove, this was a very advanced thought.

Thinking was expanding and getting closer to the mark.

He made many contributions to geometry and is credited with mathematical ideas that Isaac Newton would define many years later as the integral calculus.

It is little wonder that the Greeks thought to name a nuclear research facility after Demorcratus. He was greatly ahead of his time.

Plato (429-347 BCE) and
Aristotle (384-322 BCE)
accepted the atomos theory, and since they were so widely respected, their viewpoint held sway. Opposing views were not silenced, however.

Epicurus (341-270 BCE)
Espoused atomos with great vigor and is supposed to have written several hundred books, but none of them survived. Epicurus attracted a following, however, and one of these was a Roman known as…

Lucretius (96-? BCE).
who wrote a long poem, which survived through the Middle Ages, and which described Lucretius's views on atomos.

Now we take a long jump into the future (1500 years) while the above mentioned theories held sway and new technology would make possible advances heretofore impossible.

Pierre Gassendi (1592-1655)

was a French philosopher who read Lucretius 1500 years later, and espoused his views on atomos. Since the printing press was now well established, his books on the subject had a wide audience. For the first time the question could be posed to thousands. Prior to this point the subject of atomism could not be settled due to the inability to experimentally confirm or deny the theory. It served only as an interesting intellectual discussion that could not be resolved one way or the other. There needed to be some method of experimentation that could bring some rationale to the discussion.

What we have been dealing with to this point reflects an effort to acquire knowledge through the power of reasoning alone. The name for those who were responsible for this effort of comprehension was "philosophers" (Greek meaning lovers of wisdom).

Even in those days, philosophy took on two directions: first a turning within to attempt an understanding of human behavior, morality and ethics; second a turning out to seek explanations of other than the mind; nature to be exact. Such study, the phenomenon of nature throughout the universe, was termed natural philosophy. The word science would not make its appearance until the nineteenth century.

WHAT IS LIGHT?

The Greeks were also responsible for the first enlightened discussion on the subject of light. Prior to their serious evaluation of this phenomenon, the world was content to accept God's pronouncement: *Let there be light: and there was light.* Light was the antithesis of dark. Because of it, all life on earth was given the gift of sight. But what was it that caused our eyes to perceive the world around us when the sun or the moon or fire allowed us to see? What did our eyes do that allowed us to visualize distant objects? Did our eyes emit something that sped rapidly to a distant or nearby object, and once having struck the object caused us to see it, or did light issue forth from any luminous object and reach our eyes, and in doing so give us a visible world?

Pythagoras (582-500BCE)
championed this latter thought many years in the past.

These conflicting theories only served to raise many more questions. If light entered our eyes enabling us to see, or if something left our eyes giving us the same ability, what was it that entered or left? What about its size? What does it weigh? Very little, no doubt, if it has weight at all? What is its speed? Why does it pass through some objects and not others? Why does cloth block light and thick glass allow it to pass through?

These questions would remain unanswered for centuries, and the solution would be intimately connected with the development of the atom theory that the Greeks so brilliantly propounded.

THERE ARE ATOMS AFTER ALL

Boyle (1627-1691)
was born in Ireland with a silver spoon in his mouth. His father was the richest man in the British Isles. Boyle, to his credit, put this financial blessing to good use and took advantage of the opportunities that money opened up for him by becoming a renaissance man. The combination of brilliance and financial resources allowed him to pursue a full time "hobby" studying religion, philosophy, mathematics, languages, and the physics of such pioneers as Descartes and Galileo.

Descartes (1596-1650)
is considered the father of modern science when he refused to accept any ideas that could not be proven by experimental proof as opposed to the assumptions and emotions that served early scientists. He is remembered for his mathematical approach to physics helping to establish it as a firm discipline requiring proof as opposed to conjecture.

Galileo (1564-1642)
was an Italian scientist who concentrated his work on the physics of motion and astronomy. He postulated that the planets of the solar system did not rotate about the earth, but rather rotated around the sun raising the ire of the church in Rome who pronounced him a heretic and forced him to renounce his theories publically.

Back to Boyle

His natural philosophic work included an improved vacuum pump that allowed him to make excellent vacuums; and in so doing he demonstrated that air was necessary to sustain life, that sound would disappear in a vacuum, and a candle would stop burning as the air was evacuated.

His main triumph, so familiar to all students of chemistry, was the volume-pressure inverse relationship. Boyle used a J shaped glass tube closed at the shorter end and opened at the long end. When he poured mercury in the tube, air trapped in the closed short end. The more mercury he poured in, the less air seemed to be trapped. He made many measurements at atmospheric pressure and also at lower and higher then atmospheric pressure. He determined that when the pressure on the air was increased by the addition of more mercury, the volume of the air decreased, and when the pressure on the air was decreased the volume of air increased.

This was the pressure-volume inverse relationship and it lent credence to the atomistic theory; if air is made up of widely separated atoms, suggested by Democritus, this would explain the fact that air was lighter than solids where the atoms were closer together. Placing the atoms of air under pressure would push the atoms closer together, thus decreasing the volume of the air.

It all made sense under the atomistic theory. Others could easily reproduce Boyle's experiments, and the concept of atoms gained the upper hand, finally vindicating the Greek philosophers.

Boyle's work also laid the foundation for ending the concept of air, fire, earth, and water as the basic four elements. Boyle said speculation as to the basic elements served no purpose and an element was anything that could not be broken down further by chemical manipulation. Chemists then started on the trail of elements and eventually it was determined that gases, such as oxygen and nitrogen were elements, and iron, copper, gold, silver, and the mercury used in Boyle's experiments were elements.

This led to the definition of a compound (latin meaning put together), which was any substance put together by a combination of elements.

Joseph Louis Proust (1754-1826)
was born in France and trained as a pharmacist. He taught and did research in Spain for twenty years until Napoleon's invasion of Spain forced him back to France where he did his most important work. He was an analytical chemist and defined the law of definite proportions stating that substances only truly combine to form small numbers of compounds each of which uses components that combine in fixed proportions by weight. Chemists of the time were trying to determine the proportion of elements within each compound. Proust experimented with copper carbonate and separated it into copper, carbon, and oxygen in a ratio of five to one to four. No matter how he did the experiment this was the

result. If he added more of one of the elements in a greater proportion then there ordinarily would be in the compound, he founded some of the element left over. He found the same thing to be true for other compounds he worked with. They always had a fixed amount of elements in a definite proportion. He named this the law of definite proportions, and other chemists confirmed his work. Berthollet (see below) eventually had to admit that his concept was wrong and Proust's was right.

Proust's work supported the concept of the indivisibility of atoms.

The French chemist…
Claude Berthollet (1748-1822)
stated that the compounds could vary in composition depending upon the amount and proportion of the reactants used in the formation of the compounds.

Proust experimented with copper carbonate and separated it into copper, carbon, and oxygen in a ratio of five to one to four. No matter how he did the experiment this was the result. If he added more of one of the elements in a greater proportion then there ordinarily would be in the compound, he founded some of the element left over. He found the same thing to be true for other compounds he worked with. They always had a fixed amount of elements in a definite proportion. He named this the law of definite proportions, and other chemists confirmed his work. Berthollet eventually had to admit that his concept was wrong and Proust's was right.

Proust's work supported the concept of the indivisibility of atoms.

John Dalton (1766-1844)

was an English chemist who worked with gases and discovered that different gases could have different proportions of the same elements. For instance, the substance carbon monoxide had one part carbon and one part oxygen, while the compound carbon dioxide had one part carbon and two parts oxygen. (mon is a Greek word for one and di is a Greek word for two). He called this the law of multiple proportions, although each compound followed the law of definite proportions. He agreed that all chemical elements were composed of atoms, and each atom of an element had the same mass, whereas different elements had different atomic masses, and it was these masses that differentiated one element from another. Dalton also said that atoms combine with each other in whole number ratios. As example 1:1, 1:2, 1:3, 2:4, etc. Since atoms were indivisible, it was impossible for a fraction of an atom to combine to form a compound.

William Nicholson (1753-1815)

was an English chemist who determined that the composition of a water molecule was two parts hydrogen and one part oxygen (molecule is from the Latin meaning small mass). When he passed an electric current through water, oxygen and hydrogen bubbled out. The mass of the oxygen was eight times the mass of the two hydrogen atoms. Since there were two hydrogen atoms in every water molecule, it meant that a single oxygen atom had sixteen times more mass then a single hydrogen atom. Dalton had actually referred to this as weight, but weight refers to the gravitational pull of a body by the earth and mass refers to the amount of matter contained in the individual atoms. Mass is the preferable term, but Dalton's calling it weight has stuck and to this day we refer to atomic weights.

Jons Jacob Berzelius (1799-1848)
was a Swedish chemist who worked out atomic weights for hydrogen and many other elements. He confirmed that every element had a different atomic weight.

Dmitri Ivanovich Mendeleev (1834-1907)
was a Russian chemist who developed a table of elements with increasing atomic weights. He demonstrated that some properties of the elements repeat themselves in an orderly fashion, and these elements fall into the same column. This method of arrangement caused gaps in the columns that Mendeleev felt represented undiscovered elements. Time would prove him right.

ATOMS ELEMENTS COMPOUNDS

Boyle's work also laid the foundation for ending the concept of air, fire, earth, and water as the basic four elements. Boyle said speculation as to the basic elements served no purpose and an element was anything that could not be broken down further by chemical manipulation. Chemists then started on the trail of elements and eventually it was determined that gases, such as oxygen and nitrogen were elements, and iron, copper, gold, silver, and the mercury used in Boyle's experiments were elements.

This led to the definition of a compound (latin meaning put together), which was any substance put together by a combination of elements.

Joseph Louis Proust (1754-1826) was born in France and trained as a pharmacist. He taught and did research in Spain for twenty years. Napoleon's invasion of Spain forced him back to France where he did his most important work.

Chemists of the time were trying to determine the proportion of elements within each compound. The French chemist Claude Berthollet (1748-1822) stated that the compounds could vary in composition depending upon the amount and proportion of the reactants used in the formation of the compounds.

Proust experimented with copper carbonate and separated it into copper, carbon, and oxygen in a ratio of five to one to four. No matter how he did the experiment this was the result. If he added more of one of the elements in a greater proportion then there ordinarily would be in the compound, he founded some of the element left over. He found the same thing to be true for other compounds he worked with. They always had a fixed amount of elements in a definite proportion. He named this the law of definite proportions, and other chemists confirmed his work. Berthollet eventually had to admit that his concept was wrong and Proust's was right. Proust's work supported the concept of the indivisibility of atoms.

John Dalton (1766-1844) was an English chemist who worked with gases and discovered that different gases could have different proportions of the same elements. For instance: the substance carbon monoxide had one part carbon and one part oxygen, while the compound carbon dioxide had one part carbon and two parts oxygen. (mon is a Greek word for one and di is a Greek word for two). He called this the law of multiple proportions, although each compound followed the law of definite proportions. He agreed that all chemical elements were composed of atoms, and each atom of an element had the same mass, whereas different elements had different atomic masses, and it was these masses that differentiated one element from another. Dalton also said that atoms combine with each other in whole number ratios. As example 1:1, 1:2, 1:3, 2:4, etc. Since atoms were indivisible, it was impossible for a fraction of an atom to combine to form a compound.

William Nicholson (1753-1815) was an English chemist who determined that the composition of a water molecule was two parts hydrogen and one part oxygen (molecule is from the Latin meaning small mass). When he passed an electric current through water, oxygen and hydrogen bubbled out. The mass of the oxygen was eight times the mass of the two hydrogen atoms. Since there were two hydrogen atoms in every water molecule, it meant that a single oxygen atom had sixteen times more mass then a single hydrogen atom. Dalton had actually referred to this as weight, but weight refers to the gravitational pull of a body by the earth and mass refers to the amount of matter contained in the individual atoms. Mass is the preferable term, but Dalton's calling it weight has stuck and to this day we refer to atomic weights.

Jons Jacob Berzelius (1799-1848) was a Swedish chemist who worked out atomic weights for hydrogen and many other elements. He confirmed that every element had a different atomic weight.

Dmitri Ivanovich Mendeleev (1834-1907) was a Russian chemist who developed a table of elements with increasing atomic weights. He demonstrated that some properties of the elements repeat themselves in an orderly fashion, and these elements fall into the same column. This method of arrangement caused gaps in the columns that Mendeleev felt represented undiscovered elements. Time would prove him right.

WHAT IS THIS ATTRACTIVE FORCE?

Scientists continued to do work on the attractive force discovered by the Greeks (static electricity)...

Charles Francois Dufay (1698-1739)
was the gardener to the king of France. His restless mind moved his talents to other directions as well, and in his spare time he produced two types of electric charges when he rubbed different substances together. Working with glass, he discovered that some charged pieces would either attract or repel other charged pieces of glass. He called these two types of effects resinous and vitreous. He theorized that matter contained two fluids in a definite balance that was disturbed during the act of rubbing; each body acquiring either an excess or a deficiency of the fluids. Subsequently the terminology of resinous and vitreous for this unknown mysterious force became negative and positive. The use of the word fluid would remain for about one hundred-fifty years, or until the exact nature of this mysterious force was discovered.

In the mid-1700s,
Pieter van Musschenbroek (1692-1761)
a Dutch physicist from the University of Leyden, developed a means of storing static electricity. He half-filled a glass container with water and sealed

the top with a cork. He then pierced the cork with a nail that extended into the water. The exposed nail subjected to friction caused static electricity to be stored within the glass container. Discontinuing the friction, and touching the exposed nail, resulted in a shock. This was the earliest form of what would be known as the…

Leyden jar.
which eventually consisted of a jar coated inside and out with tin foil. The outer tin foil connected to the earth; the inner tin foil connected to a brass rod projecting through the mouth of the jar. Charging one of the tin foils with an electrostatic machine would produce a severe shock if both tin foils were touched at the same time.

It was possible to force a large amount of electric charge into the Leyden jar. In fact, if the jar carried a sufficiently large amount, even approaching a nearby object could discharge it and a spark would force its way through the air to the other object. When it did this, there were those who saw a resemblance to lightening. From this primitive beginning, natural philosophers developed numerous devices that created electric current.

Benjamin Franklin (1706-1790)
the great American politician, statesman, inventor, and natural philosopher, used a metal conductor to discharge the inner and outer layers of tin foil in a Leyden jar creating the clearly visible and audible spark. He was one of those who speculated about

the similarity of the visible and audible spark to lightening. Could lightening be an electric discharge similar to the spark produced within the Leyden jar?

To prove this hypothesis he flew a kite with a metal tip in a thunderstorm. He used wet hemp line, a conductor of electricity, to fly the kite and attached a metal key to the end of the line. From the key he attached a non-conducting silk line that he held in his hand. When he held his other hand near the key he drew sparks from it, thus proving that lightening was an electrical phenomenon with the same properties as the electric spark produced with the Leyden jar. Franklin was fortunate: the next two men who tried duplicating Franklin's experiments died of electric shock.

The lightening experiments brought great fame to Franklin. Much of his success as a diplomat in France was due to his reputation as a natural philosopher. For the French knew that they were receiving one of the world's leading scientific figures, and not just an American patriot.

Franklin showed that the electrical experiments performed in a laboratory are related to the natural events of our world---lightening. Therefore, future natural philosophic studies of nature could not be divorced from electricity.

Franklin postulated that electricity was a single fluid that existed in all matter and explained the effects of electricity by a shortage or excess of this fluid. The word fluid, used by natural philosophers, was the primitive description of a force of nature that future experimenters would gradually shed light upon…

Luigi Galvani (1737-1798),
an Italian anatomist and physician discovered that when the lower legs of a dissected frog were in proximity to an electrical source, such as a Leyden jar, the frog's muscles twitched into what he described as a "convulsive state." Galvani had uncovered an interesting phenomenon. Whereas a spark could touch a muscle and cause it to contract, Galvani demonstrated that the muscle would contract merely by being in the proximity of an electric source. Up to this point, an electric current was only demonstrated in an instantaneous fashion. Once an electric spark transferred its current, there was no more; no further current flowed. He also demonstrated that by drawing a spark from an electrical machine at a distance from a muscle, and simultaneously touching metal to the frog's sciatic nerve, the leg muscles twitched as if cramped. Further, he showed that touching a muscle with two different metals caused the muscle to twitch and the twitching would continue multiple times. Clearly, these muscles were being affected by electricity, but in this case the electricity would manifest itself not as one instantaneous jolt and then no more, but rather in a continuous fashion. How could this be? Galvani concluded that it was the muscle itself that generated the electric current responsible for this continuous twitching, and therefore animals utilized electricity as part of their physiological processes.

He named this "animal electricity," and thought that this was a different force than the natural electricity of lightening or the static electricity produced within the Leyden jar. The phenomenon of electricity was expanding in many directions.

These experiments demonstrated the electrical nature of nerve muscle function and opened the door to the study of neurophysiology.

Electrical forces in animal tissue? Did electrical energy permeate living as well as inanimate objects? Was electricity a universal force explaining life itself? These questions would "spark" much thinking and open new doors.

Alessandro Volta (1745-1827)

from Como, Italy was a student of languages and had a broad liberal arts education. Electricity became his hobby and he developed the battery in 1793. Rather than accept Galvani's notion of the muscle source for the continuous electricity, he suspected that the two dissimilar strips of metal were really the electrical generating medium. Following this hunch, he discovered that electricity was produced when two different metal strips were placed within a salt solution that had no connection with animal tissue. This simple act would produce electricity that continued as long as the chemical reaction continued. The electricity was then drawn off continuously through a wire. For the first time electricity was produced by chemical means. This was a far more powerful force that could be obtained by electrostatic machines and resulted in enhanced research efforts.

In honor of Volta's brilliant work, the *volt* now describes the unit of electrostatic potential.

It was shortly after Volta's discovery that
William Nicholson (1753-1815) working together with
Anthony Carlisle (1768-1840),
both from England, passed an electric current through water and discovered that they could break down the water into their component parts: hydrogen and oxygen. For the first time an electric current brought about a chemical reaction. They named the process hydrolysis---taking apart by electricity. Electrical insight was expanding exponentially.

Joseph Priestly (1733-1804)
of England, who emigrated to America and became a friend of Benjamin Franklin, determined that as the distance between two charged bodies of equal sign are increased by a factor, the repellant force is reduced by the square of that factor. This means that if two negatively charged pieces of metal are very close there is a strong force repelling them apart, but if they were moved, for example, two inches apart, the repelling force would be reduced by two squared—two times two or four times. Although this assumption was correct, no one demonstrated this fact experimentally. Priestly, a chemist, is better known for his discovery of oxygen.

As a result of the work by scientists mentioned above as well as many others, by the last quarter of the 18th century certain principles about electric phenomenon were established:

There are two signs of electrical charge:

Charges of the same kinds repel each other.

Charges of the opposite kinds attract each other.

Electricity was considered a kind of fluid, although there was controversy as to whether there were two types of fluid that could be added or taken away from different materials, or a single fluid that could be transferred in part or whole from one material to the next.

There are conductors of electricity through which an electric charge can freely move.

There were insulators through which no charge could move.

Electricity is stable and unchanging.

When Thales of Miletos polished amber with cat fur, and the amber was able to pick up bits of straw or feathers, there was no natural philosophic explanation for this finding. As mentioned before, what Thales had uncovered was static electricity, but this terminology would not enter the scientific lexicon for many centuries, or until further advances shed more light on the subject.

Petrus Peregrinus (1220-?)

was a French physician, who performed the first serious study of magnetism in 1269. This is probably a reflection of the fact that there was so little understood in the field of medicine, that restless minds of some physicians forced them to investigate other subjects, if for no other reason but to stimulate their minds. Working with the magnetic stone known as lodestone he described the polarity of magnets. He found that the magnetic North Pole of one magnet would be attracted to the magnetic South Pole of another. He also demonstrated that North Poles repulsed North Poles and South Poles repulsed South Poles. In short, the concept of the attraction of opposite poles and the repulsion of like poles was attributed to Peregrinus.

William Gilbert (1544-1603)

was born in Colchester, England to a family of some wealth. He went to St. John's College, Cambridge, and over a nine-year period obtained a BA, MA, and M.D. He set up medical practice in London, and held a number of offices in the Royal College of Physicians including president. He became physician to Queen Elizabeth.

As was typical of the physician scientists of the times, (see above), he held many interests, and Gilbert furthered the study of magnetism when he proposed that the entire earth was a large magnet with a North and South Pole. His book, *De Magnete,* published in 1600, became the standard upon which all work in magnetism was measured. It remains one of the classics of natural philosophic literature.

In addition, Gilbert advanced the work of Thales of Miletos, and Theophrastus when he discovered that more materials could be "electrified" when rubbed with a cloth. He called these materials "electricians" and he proposed the word "electricity," from the Latin word, *electrum*, for amber. He also theorized that electricity was due to the existence of a fluid surrounding an electrified body and the act of electrification was due to the passage of this fluid into the electrician. He also suggested that this fluid transfer was made possible when the friction of rubbing created the heat necessary to promote this passage.

Although both the rubbed amber and magnetism were attractive forces, Gilbert was the first scientist to explain the difference between magnetism and the amber effect---static electricity. The term static would not enter the physicists vocabulary for two hundred more years or until electric currents were identified as moving charges of electricity.

BEAUTIFUL LIGHTS. FORCES OF ATTRACTION AND REPULSION. MYSTERIOUS RAYS.

In medieval times, alchemy had one goal: to transform metals such as copper or lead into gold. To make "the philosopher's stone." was the dream of all natural philosophers.

Vincenzo Cascariolo (1571-1624)
of Bologna, Italy, a shoemaker and amateur alchemist, took some coal dust and barium sulfate, heated the mixture and spread it over a piece of iron. Now he would sit back and reap his harvest. The next day he discovered a phenomenon which in time would be more valuable than his failed attempt to transmute a base metal into gold: the treated metal glowed. He sat in awe at what he had accomplished and felt sure that he had taken the first step in the process of transmutation. To Cascariolo's disappointment, the glow soon faded, but he was able to bring it back by placing the bar in the sunlight for several hours.

The process of a substance, or luminescent material, giving off a glowing light when exposed to an external source such as the sun's rays is known as fluorescence. Once the external source stops, the glow ends immediately.

When a luminescent material continues glowing for some time after the external source ends, that effect is called phosphorescence.

In the primitive natural philosophic knowledge of the day, all that could be gleamed from Cascariolo's discovery was that the stone captured the sun's rays. "The Sun Stone" found use as covering for religious icons, crucifixes, and rosary beads. It soon became the thinking of the time that prayers said in the presence of these glowing icons would be answered much sooner. This is an example of how some discoveries, made before science has evolved enough to explain it, takes on religious or pseudoscientific overtones.

Otto von Guericke (1602-1686)
received his education at the University of Leipzig, Germany. He also studied law at the University of Jena, and mathematics and mechanics at the University of Leyden; another reflection of renaissance-man thinking.

As a trained engineer, he took a ball of sulfur, attached it to a hand crank and when he rotated it against his hand, it produced an electric charge. He had developed the first electric generator. He also ascertained that the electricity would cause the sulfur ball to glow, thus discovering that electricity could cause fluorescence. Whereas Cascariolo's energy source for this phenomenon was the sun, von Guericke's was an electric current. Natural philosophers would eventually discover other energy sources to create fluorescence.

Von Guericke also transmitted an electric charge through an ordinary piece of thread. Of course this was too early for any person of the time who discovered this phenomenon to realize that he had developed the forerunner of electric circuitry. In addition, von Guericke invented the air pump. With this, he was able to create a partial vacuum. He showed that light will travel through a vacuum but sound could not. Example: the sun is visible, but we can't hear it as the sun's rays pass through the vacuum of space.

The information about light being able to travel through a vacuum would have great future relevance.

Von Guericke took two copper bowls, placed them together to form a sphere, and evacuated the air between them. To the amazement of onlookers, a team of eight horses, tethered to a rope attached to each sphere, and pulling in opposite directions, could not pull the bowls apart. The conclusion from this phenomenon was that air pressure alone was responsible for holding the bowls together, and surprisingly was stronger than the power of the horses.

Von Guericke's air pump would have great relevance for future advances in the field of electricity and physics. Early discoverers of such phenomenon could not possible have the insight to visualize future ramifications. It would take, others acting on the pioneers advances and future developments, to advance thinking in enhanced directions.

Stephen Gray (1666-1736)

was the son of a dyer from Canterbury, England. Working with substances electrified by static electricity, he discovered that using certain materials would allow the charge to spread from that body (future electrical conductors). Other materials tested would not exhibit the same property (future electrical insulators).

Charles Coulomb (1736-1806)

went to military school in France where he studied engineering. It fell to him to refine Priestly's thesis that as the distance between two charged bodies of equal sign are increased by a factor, the repellant force is reduced by the square of that factor. This means that if two negatively charged pieces of metal are very close there is a strong force repelling them apart, but if they were moved, for example, two inches apart, the repelling force would be reduced by two squared—two times two, or four times. Through experimental methods, he was able to demonstrate this inverse square law. He published the result, and received full credit for the concept now known as Coulomb's law. A unit of electric charge (coulomb) became known as a unit of energy.

He also refuted Franklin's single-fluid theory, promoting the two-fluid theory of electrical charges.

Hans Christian Oersted (1777-1851),

a Danish physicist, was aware of a long understood but never studied phenomenon that lightening would cause a compass needle to change position. He believed that this simple observation unified the laws of nature and he set out to prove that electricity and magnetism were in some way related. While Oersted lectured his students about electric theory, he discovered that whenever the electric current was turned on, a nearby compass needle moved. Although these forces were considered separate, Oersted demonstrated their intimate partnership. At the time, of course, the significance of this phenomenon could not be visualized, but this opened the door to future great unification discoveries between electricity and magnetism. The oersted is a unit of magnetic intensity.

Dominique Francoise Arago (1786-1853),

a Frenchman, demonstrated that a wire carrying an electric current would show magnetic properties and attract iron filings just as a magnet did. This attractive force ceased immediately when the current stopped. To this point, magnetism was possible only with iron, but now Arago demonstrated that an electrical current flowing through copper wire could also act as a magnet.

The work of Oersted and Arago demonstrated that an electric force could create magnetism. Scientists now began to think of a unified electromagnetic force: electromagnetism.

Andre Marie Ampere (1775-1836)

was a French physicist who discovered that if an electric current flowed through two parallel wires, an attractive force set up between them. On the other hand, if the current through these two wires ran in opposite directions, a repulsive force developed between them. On the basis of the inverse square laws, he worked out the mathematical relationship between the forces in these wires.

Ampere also twisted the wires in the shape of a coil and sent a current of electricity through the coil. This caused formation of a magnetic field, and the proximity of these magnetic fields would cause them to reinforce each other. The resultant effect was that the coil of wire with its current would act like a bar magnet. The results of these experiments were two-fold and led to the development of electromagnets and a greater understanding of the interaction between electricity and magnetism. The ampere became the unit of electric current.

Michael Faraday (1791-1867)

is considered one of the great scientific geniuses of all time. This fact becomes all the more amazing when one realizes that he was self-educated, never attended a university, and received only a grade school education. Einstein kept a picture of him in his office.

Born in England, his parents were devout Sandamanians, a Protestant sect believing strongly in physical law. Faraday was deeply influenced by the combination of strong moral teachings and the belief that the dedicated pursuit of physical law represented the discovery of God's unifying plan, thus honoring God.

His family was extremely poor, and Faraday's education consisted of "the rudiments of reading writing, and arithmetic." At the age of fourteen he was apprenticed to a bookbinder and besides learning a trade he became exposed to many scientific books of the time. Anxious to take advantage of this opportunity, he diligently improved his reading skills. He set a goal on becoming a natural philosopher, and at age twenty-one took the unprecedented step of writing to Humphry Davy, a world famous chemist and president of the Royal Institution. On the strength of his letter, Davy gave Faraday free tickets to attend the Davy lecture series. Faraday took copious notes, bound them in book form and presented the book to Davy himself. This bold step resulted in a position at the Royal Institution under Davy's tutelage.

Gradually his reputation spread as he published many scientific papers. Within twelve years, he became a fellow of the Royal Society and director of the Royal Institution.

As part of his religious orientation, he felt that gravity, electricity, and magnetism could be unified into a single force, but efforts to prove this were to no avail. This concept, still considered the Holy Grail, put physicists on a continuing search for grand unification theories to this day.

What Faraday's predecessors established, however, was an ill-defined connection between electricity and magnetism. Faraday attempted a clarification and after years of experimentation, he finally succeeded in unifying these two forces. He took a doughnut shaped magnet and wrapped one side with wire attached to a battery. He wrapped the other side of the doughnut and attached the wire to a voltmeter. When he sent current through the first wire, the voltmeter needle moved and then returned to its original position and stayed there. When Faraday turned off the voltage into the first wire, the voltmeter needle again moved. An astonished Faraday kept turning the current on and off and watched in amazement as the voltmeter needle moved only during the on-off motion. Clearly, the electricity had induced a magnetic force in the doughnut magnet which then induced an electric current in the second coil of wire. At the age of forty, he reported the historic connection in plain English as opposed to the usual mathematical language of most scientists: the increase or decrease of a magnetic force produces electricity. A stable, unchanging magnetic force produces none. The amount of electricity produced depends upon the speed of the increase or decrease. In spite of the fact that Faraday did not utilize the language of mathematics to describe the relationship, his colleagues immediately recognized the brilliance of his work and accepted his findings. The simple truth was that Faraday was not capable of utilizing mathematics since he never had the opportunity to study the subject. To his incredible credit, this deficiency never held him back!

The absence of mathematics allowed Faraday to verbalize what he called "tubes of force." Here his religious upbringing led him to the conclusion that he received a God given gift allowing him to explain the magnetic and electrical effects of the wire and metal. But even more importantly, he developed the thesis that the magnet sent out "tubes of force" in all directions and encompassed the entire world in its grasp. The science of electromagnetism was born. This was the first step taken in the quest to unify the forces of nature.

The efforts of Faraday resulted in the dynamo and the electric motor. Civilization had entered the electric age.

In what would become one of the great universal theories in the history of physics, in a series of four equations,

James Clerk Maxwell (1831-1874),
a Scottish physicist, formulated the mathematics of Faraday's experimentally observed relationship between electricity and magnetism. In other words, he placed Faraday's plain English on a firm mathematical footing and proved Faraday's "tubes of force." The field theory had been born. Faraday's "tubes" had become the all-encompassing electromagnetic field.

Here are Maxwell's equations:

$$\nabla \cdot \mathbf{E} = 4\pi\rho$$

$$\nabla \times \mathbf{E} = -\frac{1}{c}\frac{\partial \mathbf{B}}{\partial t}$$

$$\nabla \cdot \mathbf{B} = 0$$

$$\nabla \times \mathbf{B} = \frac{4\pi}{c}\mathbf{J} + \frac{1}{c}\frac{\partial \mathbf{E}}{\partial t},$$

don't ask me to explain them

These four equations led to his prediction of electromagnetic waves that traveled at the speed of light, and along with Newtonian Mechanics became the foundation stone upon which all physics depended. Indeed, in Maxwell's mind and now in the mind of most physicists, light was a wave…an electromagnetic wave. It was not a particle. Could it be that the ancient Greek philosopher's questions were about to be answered? Maxwell's electromagnetic theories were proven within twenty years.

Maxwell's brilliant mathematical insight would bring him into competition with Einstein and Newton (more about these two men later) for the greatest intellectual achievement of all time. In fact, prior to Einstein's work, Maxwell had joined Newton in forming a double underpinning for all of physics: Newtonian Mechanics, and Maxwell's Electrodynamics (Classical Physics). Now, physicists felt they had an orderly, simplified picture of the laws of nature that explained the physical realities of the world requiring only clarification of a few more decimal points. Some even contemplated switching to another field, because there was nothing left to contemplate in physics.

But as we have seen, and as we shall continue to see, nature objected to being cast in stone. Soon various experimental and theoretical physicists would come upon a series of mysterious discoveries that no one could explain; such things as unusual peculiarities of light radiated by black bodies, strange rays labeled x-rays because no one knew what they were, the phenomenon of radioactivity where mysterious rays emanated from the atom itself without external stimulation, the photoelectric effect or electricity induced by light itself. Physicists would soon come upon an unfamiliar New World that would change the way they thought about their fascinating discipline.

Now that there was a chemical battery invented by Volta, numerous advances in design enabled the battery to produce ever increasingly powerful electric currents. This made it possible to move electric currents across a vacuum, so physicists attempted the development of more and more powerful vacuums. In 1855,

Heinrich Geissler (1815-1879)
a German physicist, developed a mercury pump, which greatly improved the ability to make a vacuum. He was able to evacuate air from a glass

container by trapping air with rising and falling levels of mercury and produced a vacuum with only tenth of one percent of the air left in the glass container.

Then he embedded wires in both ends of the vacuum tube, and passed an electric current from the negative electrode, known as the cathode, through the vacuum, and toward the positive electrode, known as the anode, at the other end. He noted colorful glows of light within the tube. These glows or "rays" emanating from the cathode would reach the glass by the anode and produce a fluorescent glow. What were these rays? Could they be traveling in the invisible medium called the aether? At the time physicists thought the aether necessary to allow light waves to travel through space. Could these rays be similar to light? Were the rays a particle of matter? What was the relationship between electricity and matter? What caused the production of these rays? Would evacuated glass tubes filled with rarified gases, through which an electric current passed, open the door to the atom and the secret of matter? The questions posed by the ancient Greek philosophers were on the threshold of resolution.

Julius Plucker (1781-1840)

a German mathematician and physicist was a friend of Geissler's and he named his vacuum tubes Geissler tubes. Plucker first placed the two pieces of wire into opposite ends of the tubes and connected them to the cathode and the anode thus making possible the study of electric currents through a vacuum. When he made this connection and caused electricity to flow, he noted a green luminescence originating from the cathode. It never varied regardless of what metal he used to make the cathode or what minute quantities of gas was left in the tube after the vacuum was created. Plucker also demonstrated that a magnet would shift the luminescence to one direction or the other depending on whether he used the North Magnetic Pole or the South Magnetic Pole. Clearly, this confirmed that the luminescence was an electrical phenomenon since the connection between the two forces of magnetism and electricity was already established. Further experimentation demonstrated that the luminescence was not only present at the cathode, but also could reach the other side by the anode. Interestingly enough, if he moved the anode, the rays still traveled in a straight line, missed the anode and struck the glass on the other side causing the same green luminescence.

Johann Wilhelm Hittorf (1824-1914)
was a German physicist who demonstrated that if one placed a solid object in front of the cathode where the rays originated, there would be a shadow visible on the opposite anode end of the tube. What was traveling from the cathode and stopped by the metal obstacle placed in its path?

He also demonstrated that a magnet could deflect these mysterious rays. What was the connection between these rays and magnetism?

Hittorf and Plucker

established that these mysterious rays traveled in a straight line and could cast a shadow if an obstruction was placed in their path.

As is so true in science, every advance throughout the ages raises more questions.

Eugen Goldstein (1850-1930),

a German physicist, took note of the origin of these mysterious rays and he called the electric energy source "cathode rays." He demonstrated that these cathode rays emitted perpendicular from the surface of the cathode.

Physicists were now embroiled in a controversy as to the exact nature of these cathode rays. Since they traveled in a straight line and were not influenced by gravitational attraction, some assumed this meant they were waves. Others felt that the rays were electrically charged particles since a magnet affected them. Those that made this assumption dealt with the fact that the particles were unaffected by gravity simply because they moved too rapidly and their mass was too small. This set up a wave particle controversy that would persist for many years or until a compromise could be reached.

Goldstein performed another interesting experiment. He used a perforated cathode in place of a solid cathode. He noted that at the same time the electric potential caused negatively charged cathode rays to speed toward the anode, another ray went through the perforations in the cathode in the opposite direction from the anode. Since they went through the perforations, he called them canal rays. It was safe, however, to conclude that since the canal rays moved in the opposite direction from the negatively charged cathode rays, they must be positively charged.

These would turn out to be alpha particles of which we will later write more.

For the next forty years, physicists would be occupied attempting to determine the nature of these mysterious rays. As we will learn, they were uncovering atomic secrets one by one.

Goldstein's last paper was published in 1928, two years before his death. He discharged electricity through a glass tube containing nitrogen and hydrogen. To his surprise he found traces of ammonia in the tube. Clearly the electric discharge had promoted the merging of the two elements. Many years later scientists investigated the possibility that life on earth could have started this way.

William Crookes (1832-1919)
of England was a chemist and physicist who could live every man's dream. He inherited a financial fortune from his father, thus enabling him to pursue his goal of pure science. From age twenty-four on, he devoted himself to scientific work in his own private laboratory. Crookes developed an even more powerful device for making a vacuum, eliminating more than ninety-nine percent of the air remaining in Geissler's tubes. These Crookes tubes used an

even higher voltage, thus making possible improved observation of the rays emanating from the cathode ray tube. They were now thinner and more clearly defined. He also showed that cathode rays traveled in straight lines and produced phosphorescence when they struck certain materials. He also demonstrated that these rays were energetic enough to turn a small paddle wheel placed in their path.

These findings put Crookes on the side of the physicists who believed that the rays were of a corpuscular (particle), or physical nature as opposed to waves.

On the strength of these advances, atomic and electrical nature of matter became well established. Still unexplained was the connection between these forces. The investigators were anxious to prove this essential link and the pace of research accelerated.

Water waves require the media of water through which to travel, and sound waves require the media of air. It was therefore believed that electromagnetic waves required a media as well, and, as we have discussed, the aether, or luminiferous aether was the media through which electromagnetic waves traveled, or so the physicists of the world thought. This was not a new concept, but had been a given in physics since the Greek philosophers. Sound waves, for instance, generate by a vibrating object such as a tuning fork or bell or vocal cords, and send out waves that alternately push and pull on the air, that is---compress and decompress. When the waves reach the ear, the eardrum is alternately pushed and pulled with the same frequency---pushing and pulling---per second as the sound. This action in turn stimulates auditory nerves, which transmit the frequency to the brain. We then hear the loudness or intensity, and pitch or frequency. In water, we clearly see the ripples caused by any disturbance of the water. These ripples form waves on the water's surface.

The aether was born of necessity to explain what waves traveling in space pushed against. It was the space equivalent of the water and air necessary as media for water waves and sound waves. Whatever the aether was, it was clearly invisible, could not be touched or smelled, and caused no interference with the waves propagated or the particles racing through space. It was an accepted concept in spite of the fact that a light wave, or an electromagnetic wave, reached us from the sun and stars, and traveled through the vacuum of space. To explain the nature of waves and the requirement that waves, in order to be waves, had to have something to push against, physicists postulated the aether and considered it to permeate the atmosphere and the vacuum of space. How else would we be able to see the sun and the stars if the waves had no media to travel through?

No one, however, could prove that the aether existed, but two physicists from America, finally relegated the aether to an interesting footnote in physics history. The speed of sound varies with the force of the wind. Faster if it travels with the wind and slower if it travels against the wind. It was thought that the same thing must be true with light.

Therefore, its speed will vary if light travels with the aether or travels against it; or so it was thought.

Albert Michelson (1853-1931)
Edward Morley (1838-1923)
two American physicists, in an ingenious set of experiments performed in 1887, conclusively proved that regardless of the way the aether moved with relation to the moving earth, the speed of light never varied (as does sound) and remained constant. The reluctant conclusion was clearly that the whole concept of the aether had to be wrong. But if there was no aether how could light waves propogate? The answer would have to await the work of Einstein eighteen years later.

Heinrich Daniel Ruhmkorff (1803-1877)
a German engineer, developed an induction coil that could produce long sparks of electricity. This was a much more powerful device used by physicists in their matter-electricity experiments. He wrapped two coils, insulated from each other, around a cylindrical iron core. The first coil was made of thick wire turned around the core a few times. The second coil was made of many miles of thin wire also turned around the coil. A battery produced a current in the first coil and when a breaker interrupted this current, a strong current induced in the secondary coil then gives off the spark. Ruhmkorff clearly developed a practical use of Faraday's great discovery. His machines would play

a significant part in the advances of electric theory and physics.

Heinrich Hertz (1857-1894),
a German physicist, the son of a prominent lawyer and legislator, was the first to produce electromagnetic waves in the laboratory and he measured their length and velocity. Before we discuss his experiments, it will be helpful to quickly review the concepts of James Clerk Maxwell.

Maxwell clarified mathematically the closeness of electricity and magnetism thus promoting the concept of a single electromagnetic radiation. He proved that a flow of electricity changed periodically from a maximum to a minimum. This change, known as oscillation, should produce a flow of electromagnetic radiation traveling with the speed of light. From this arose the certainty that light itself was an electromagnetic radiation, for since all electromagnetic radiation traveled with the speed of light it would be far-fetched not to assume that light was not also an electromagnetic radiation. If correct, this would mean that it would be theoretically possible for man to produce an electromagnetic radiation by oscillating an electric current. How to prove this was another mystery, for man was incapable of oscillating electricity fast enough to produce tiny wavelengths of light. This would have required more than one quadrillion oscillations per second (1,000,000,000,000,000). However, it might be possible to produce about one thousand oscillations per second. And since the oscillations are traveling at the speed of light---three hundred thousand kilometers per second, an individual wave-length would be about 300 kilometers long. According to Maxwell, these long waves should also exist.

Hertz developed an oscillating circuit with two metal rods placed end to end with a gap between them. When charges of opposite signs were sent through these rods, strong enough to spark in the gap, the current would oscillate back and forth across the gap and also in the rods. He then detected the electromagnetic waves liberated with the use of a nearby similar circuit. What he had learned was that the electric and magnetic fields could detach themselves from the wire and roam freely through space as Maxwell's electromagnetic waves.

Hertz was a pure scientist and made this discovery in the name of pure science. When asked about potential uses of these waves, he replied that they were of no use whatsoever. He never thought of any potential commercial possibilities, interested only in discovery for the sake of furthering knowledge. It took…

Guglielmo Marconi (1874-1937)

an Italian electrical engineer to turn Hertz's discovery into the radio. Hertz confirmed Maxwell, and discovered radio waves, an invisible part of the electromagnetic spectrum.

Unfortunately for science, Hertz died of septicemia (blood poisoning) at the age of thirty-seven…

Phillip Lenard (1863-1947)

was a prominent German physicist who was frustrated by the fact that the cathode rays were only studied through a glass tube. He took note of the work of Heinrich Hertz who also established that the rays could pass through aluminum foil. Lenard had to roll out the foil to twenty five thousandth of an inch before the rays could pass through. From this he concluded that they must be smaller than atoms. Also, since the rays produce a quantity of heat when they strike a solid glass wall, they must be traveling at enormous speeds.

The researchers were zeroing in on their target, but the exact nature of the rays remained elusive.

Lenard was also one of those physicists who would object to the dismissal of the concept of the aether in physics. He became frustrated enough, and lived long enough to usher in the Nazi era, become devoted to the cause, and espouse the concept of the decadence of "Jewish Physics," the soon to be developed quantum mechanics and relativity, Against these concepts, he promoted the supremacy of "Aryan Physics."

In the last one hundred years physics had advanced more than it had from the time of Thales in 600BC to 1800 AD. However, starting in the late nineteenth century and early twentieth century, the greatest advances were yet to come.

Many scientists made advances in the study of the phenomenon of cathode rays developed in a rarified glass container by electricity generated by Ruhmkorff's induction coil. These scientists

gathered results that put them on the path of further inquiry and discovery. Cathode rays, whatever they were, exhibited certain characteristics:
- They traveled as straight as an arrow.
- They clearly arose from the negative cathode and carried a negative charge.
- Since they could pass through some substances placed in their path, they had to be very small or wave like.
- When these rays struck the far glass wall of the cathode ray tube, they produced heat. If a wave or particle of such infinitesimally tiny size could impart energy in the form of heat, this meant that they had to be traveling at enormous speed.
- A cathode ray, deflected by magnetism, suggested that these forces exerted an influence on each other.
- The cathode ray could push a paddlewheel up an incline confirming that the rays had energy and could produce work.

THE FIRST LOOK INSIDE THE ATOM...X RAYS

No advance had such sudden impact as the discovery by...

Wilhelm Konrad Roentgen (1845-1923),
a German physicist who was educated in the Netherlands where he lived from the age of three. He received his doctorate in physics at the University of Zurich.

One evening while studying the phenomenon of cathode rays, he noted that a powdery luminescent substance suddenly began to give off a light in the darkness. The luminescent substance was lying near a rarified glass container in a tightly sealed black carton that Roentgen had prepared in order to test the transparency of various types of paper to the cathode rays. Clearly, the luminescent powder was glowing while under the influence of something emanating from the electrified and rarified glass tube. He took a standard laboratory piece of equipment, a screen, which was a paper plate with one side covered with a layer of luminescent barium platinocyanide. He moved it next to the glass tube. It immediately fluoresced and continued to do so even as far as five feet away. What were these mysterious emanations that went through the glass tube and into the room? Roentgen unknowingly had opened the door into the atom and saw what came out. Just as no one knew the exact nature of cathode rays, Roentgen was in the dark about the nature of his mysterious emanations so he named them x-rays. Subsequently, Roentgen established the permeability of such rays through various substances. Some were completely permeable like paper, others showed varying degrees of permeability, and lead was impermeable. Once when Roentgen turned the tube, his hand was interspersed between the tube and the luminescent screen and for an instant he caught a fleeting impression of the bones of his hands on the screen. Roentgen took the first recorded x-ray in history when he placed his wife's hand in the path of the rays. When developed, the plate showed her soft tissues, bones, and wedding ring. His discovery revolutionized diagnosis in medicine.

Cathode rays, however, were still unexplained.

A FURTHER LOOK INSIDE THE ATOM...RADIOACTIVITY

Antoine Henri Becquerel (1852-1908)
of France was destined to become a scientist. His grandfather contributed to the field of electrochemistry and his father to the phenomenon of fluorescence and phosphorescence. Becquerel, fascinated by the work of Roentgen, questioned the origin of x-rays. Could it be that the actual origin of the x-rays was in the materials used to make luminescent screens?

For the most part the substances used as coating on luminescent screens were uranium salts. He took crystals of double sulfate of uranium and potassium, a substance known to fluoresce upon exposure to outside energy such as the ultraviolet radiation of the sun, and placed them on a photographic plate covered by sheets of black paper. He put the crystals on top of the black paper that covered the photographic plate. Sunlight could not penetrate the black paper, and any fluorescence the crystals gave off would likewise not penetrate the black paper. However, if the crystals gave off x-rays, then the black paper would be easily penetrated and blacken the photographic paper. After sun exposure and development of the photographic paper, Becquerel did indeed find the photographic paper blackened. An excited Becquerel was convinced that the uranium, if exposed to sunlight, emitted Roentgen's mysterious x-rays. He reported his findings to the French Academy of Sciences.

But he soon had to retract. One day, Becquerel prepared the same experiment with the uranium salts, but because there was no sunlight, he placed the preparation in a drawer where there were also some undeveloped photographic screens. Several days later, when the sun shone forth, he attempted to repeat his experiments. First, he developed the photographic screens to make sure the uranium salts lying nearby had not affected them. Then when he was sure the photographic screens had not been affected, he would use them in his experiment. He was shocked to discover that fluorescence was intensely manifested in black on the screen in spite of the absence of sunlight.

Becquerel realized that this had to be a spontaneous fluorescence by the uranium itself void of the usual outside stimulating influences like the ultraviolet rays of the sun. He reported this new finding to the academy, stating that this was a radiation similar to Roentgens in that it could penetrate opaque bodies, and could darken a photographic plate enclosed in a lightproof container. He cautiously stated that, in spite of these similar characteristics, he could not say for certain if his rays were x-rays because the place of origin was not a cathode ray tube but rather the uranium salts themselves.

It was immediately grasped that this accomplishment was more than an incidental discovery, but had pronounced significance.

Physicists all over the world began serious study of this phenomenon. In time, their study would result in revolutionary advances about the atomic structure of matter.

A young Polish physicist,

Marie Curie (1867-1934)
nee Maria Sklodowska, did her doctoral thesis on Becquerel's phenomenon. She decided to determine if there were other elements that would exhibit the same properties as uranium. Working with the mineral pitchblende, she succeeded in isolating the active ingredient responsible for this activity: radium. In addition, she identified a new substance she named polonium in honor of her native land, Poland. Becquerel and Marie Curie and her husband...

Pierre Curie (1859-1906)
opened the door to a new science named radioactivity that in turn would result in future developments in atomic physics.

By now, physicists began to understand that radioactivity was an astonishing phenomenon. They had learned that the energy liberated by radioactivity in the uranium atom was enormous, but in radium, they calculated that the energy liberated was several million times more. And think of this: uranium liberates alpha particles (proven later to be a helium atom) continuously, and these alpha particles travel at ten percent of the speed of light or eighteen thousand six hundred miles per second. Uranium also liberates so-called beta particles, and these travel at ninety percent of the speed of light or one hundred sixty seven thousand four hundred miles per second. What were these beta particles?

When you hear the ad on television, which tells you that car x accelerates to sixty miles per hour in six seconds from a standing start, the advertisers are trying to impress you with the automobiles tremendous power. Imagine then, the power that must be responsible for accelerating a particle instantly to eighteen thousand miles per second from a standing start. Why not use the example of the beta particle traveling at ninety percent of the speed of light? Why? Because, as you will learn later, beta particles aren't ejected from a standing start.

In both uranium and radium, the energy liberated was not only unusually high, but continued indefinitely: hour after hour, year after year and century after century. Where in heaven's name did all that energy come from? Could there be so much energy tied up in the nucleus of the atom? It seemed as if energy was being created out of nothing, but clearly, this violated the law of conservation of energy which said that the energy in any system was constant and therefore energy could not be created out of thin air. But, was it the case with radioactivity? Perhaps then, it seemed more logical to conclude that a radioactive atom must absorb energy from its environment and then convert it into another form of energy in spite of what Becquerel or the Curies postulated. However, most physicists accepted the fact that radioactivity meant that all atoms had this great amount of energy trapped within their structure. The fact that physicists had no way of releasing this trapped energy was the reason it had remained unknown. But now, the concept of radioactivity had demonstrated that this massive amount of energy was there and its slow radioactive release provided the clue. As we will see later, it would take Einstein to finally clarify this riddle for his colleagues.

Working under primitive conditions, and not being aware of the dangers of radioactivity, Madame Curie paid the ultimate price, dying of aplastic anemia undoubtedly induced by the radiation.

BACK TO LIGHT

In the nineteenth century when electric current was passed through the glass containers filled with rarified air or gases, the scientists were both enamored of the beauty of the colorful displays and were also certain that they had clarified the physical properties of the light shining within. Light not only was produced within the glass tubes, but permeated our world. The light from the sun made it possible for us to see.

This, of course, was not enough of an explanation for curious scientists. What is light? Does it have a shape? How do our eyes and brain perceive light?

Thomas Young (1773-1829)
As mentioned, Thomas Young, a precocious youngster, was an English physicist who had learned nine languages including mathematics by the time he was sixteen years old. He became a scientist, and in keeping with the renaissance-man thinking of the time also became a physician. Prior to his arrival on the scene, Isaac Newton's particle theory of light was the accepted concept, but Young dared to overturn the thinking of this giant and declared that light was a wave. Just as a ripple forms on the water when a pebble drops therein, so do light waves travel through space.

Young took a board with two slits and passed light through one slit at a time. When the light waves hit a wall behind the slit, and a graph of the intensity of the light on the wall plotted, it could be demonstrated that the light waves were most intense directly opposite the slit. The light waves that hit the wall farther away from the slit declined in intensity proportional to the distance from the slit. If Young passed the light through both slits, the intensity of the light that shone on the wall peaked and troughed in many different places on the wall. Clearly, the waves coming from both slits had crossed each other and in so doing they both combined and strengthened their intensity or canceled each other out thus reducing the intensity down to zero. This "interference pattern" similar to what happens when two ripples on the water intersect was diagnostic for wave phenomenon and therefore it had been proven conclusively, or so Young thought, that light was a wave.

The scientific community, on the other hand, was aghast at the thought of this upstart daring to contradict Newton and they declared that his hypothesis "lacked any trace of learning." After all, Newtonian Mechanics and the discovery of gravity were the pillars upon which they built physics.

This caused Young to turn to other interests, so he decided to assist in deciphering the Rosetta Stone. Soon, however, more and more physicists embraced the wave concept, and it became sacrosanct as was Newton's corpuscular concept before Young.

THE UNIVERSE IN MOTION

Newtonian Mechanics was so critically important in the development of physics that it behooves us to spend some time explaining it. And in order to understand what a giant leap forward it represented, a historical summary is in order.

The key to understanding motion was critical to the comprehension of how the world and the universe worked. With no motion, there would be no world or universe. Motion is tied to space and time. Something moves over a distance in a set period of time. Physicists understood that there could not be any spontaneous motion. What is standing here is here and what is here can be there only over a period of time and after having traveled through some distance. The ancients wrestled with the question as to whether there was a natural state of motion.

On the basis of visual observation of the world and the heavens, the ancients thought there were two motions: one was earthly and the second was heavenly, and as a result they came up with two sets of rules for these different realms.

It seemed to be that motion of a body on the earth was always in a direction pointing toward the center of the earth. If I drop a ball or throw a ball in the air, it will seek the center of the earth. This is the eventual direction it will take and it will take it whether you are standing in England or China or the North or South Pole. The body in motion, of course, will end up at rest in its path to the earth's center as it hits the ground or lands on any other object in its path. The principle that must be understood is that the object in motion will try to get as close to the center of the earth as it can.

But the heavenly realm was different, because as the ancients viewed the heavens and followed the course of the stars day after day and month after month it was clear that they took a circular path. Therefore, the natural motion of the heavens was circular motion.

With these principles in mind, we come to Aristotle the first person to establish a picture of our universe.

Aristotle (384-322 BCE)
As we have learned previously, early "thinkers" postulated that the earth was at the center of the universe. The sun and stars and planets moved around the earth in the circular motion so clearly visible through observation of the heavens. There was, however, an exception to this neat picture, and that was that certain celestial bodies, the planets, did not move in a perfect circle around the earth, but on occasion would seem to pause and actually reverse their motion. So another natural philosopher of the day came up with a slightly altered view of the heavenly realm.

Ptolemy (100-178)
modified Aristotle's view and suggested that the planets as they circled about the earth moved in

little circles around their larger circular orbiting path. The theory put together by Ptolemy and Aristotle would reign for 1500 years.

Copernicus (1473-1543)
came up with a radical departure from Ptolemy's model when he said that it was the sun at the center, and circling around the sun were the planets making their little circles as they rotated. This heliocentric or sun centered view caused quite a stir, especially amongst the Catholic religious, for anything that did not enhance the primacy of earth and mankind was vilified.

Johannes Kepler (1571-1630)
studying the planetary measurements of...

Tycho Brahe (1546-1601),
a famous Danish astronomer, decided that the planets actually traveled around the sun in ellipses rather than circles.

Through all the 1500 years of these theories there was still a dichotomy between motion in the earthly realm and motion in the heavenly realm. But then another great scientist added his insight...

Gallileo Galilei (1564-1642)
through his invention of the telescope observed many phenomena that suggested that the heavenly realm may not be so different from the earthly realm. For instance: the four moons revolving around Jupiter like a mini solar system; the phases

of Venus demonstrating a smaller size when full (on the other side of the sun) and larger when crescent (on the same side of the sun as the earth). This confirmed the heliocentric view. The mountains and craters on the moon suggesting a similarity to earth also suggested a similarity with the earthly realm.

In addition, through experimental observation, he decided that an object in motion would always travel in a straight line forever if there were no friction and air resistance. This upset the notion of Aristotle that claimed it was natural for an object to be at rest and it moved only if pushed or pulled by an external force. Instead, Galiliei said that all objects have inertia and the natural tendency is to keep moving, and the object wants to move in a straight line and at constant speed; that is, it has uniform motion. This concept was difficult to grasp for intuitively and observationally the natural state seems to be that objects are at rest on the earth's surface and do not move unless exposed to an external force.

The concept of normal motion being straight-line uniform motion and not a resting state was a giant leap forward, developed at a time when it seemed counterintuitive. And this is a characteristic of great thinkers; they have no fear of upsetting established dogma.

Then another great scientist came forward and took the ultimate leap into one of the great insights of all time.

Isaac Newton (1642-1727)

"If I have seen further than others it is because I have stood on the shoulders of giants."

Newton agreed with Gallileo and stated that an object's natural motion is to move uniformly in a straight line unless acted upon by an external force. This became known as the law of inertia. Second, he stated that the force on any object is equal to its mass times its acceleration (F=MA). In other words, a larger force produces a larger acceleration, and it will take more force to move a more massive object. Third, he stated that for every action there is an equal and opposite reaction. Example: the force of a bullet leaving a rifle's chamber causes the rifle to recoil.

Then came his greatest insight when, as the story is told, the apple fell from the tree.

He developed his theory of gravity. He realized, most likely by thought experiment, that the falling apple was no different than the moon circling around the earth. As the apple fell, or accelerated toward the center of the earth because of gravity, so indeed was the moon falling in space around the earth. As the moon was in straight line uniform motion, the earth's gravitational pull acted as the external force changing the moon's natural straight line according to Newton's first law. Indeed if the gravitational pull of the earth on the moon suddenly disappeared, the moon would be shot off from its orbit in a straight-line path just as a ball on a string would if it was twirled around one's head and the string suddenly broke.

By his universal law of gravitation, Newton was the first to unify the earthly and heavenly realms. Both of these realms had to follow the same physical laws of motion. This represented one of the greatest intellectual achievements of all time. And Newton added another achievement when he invented the calculus. With this new mathematical construct, he was able to calculate, for instance, how gravitation causes elliptical orbits.

Put another way, the laws of motion are the same for everyone in uniform motion whether that motion be in a space ship going nearly at the speed of light or in an airplane, or in an automobile on the surface of the earth.

The laws of motion, as already stated, are the same for everyone in uniform motion and it is not enough to say, "I am going fifty miles per hour." That is a meaningless statement. Rather one would have to say I am going fifty miles per hour relative to the street. And if you travel fifty miles per hour in a car and pass a pedestrian standing on the sidewalk, are you going fifty miles per hour or are you at rest and watching through your rear view mirror as the pedestrian leaves you at fifty miles per hour?

All this, given the name of Gallilean relativity, would be the forerunner of another theory of relativity to come hundreds of years later.

Again, the experiments of Michael Faraday and James Clerk Maxwell clarified the close relationship between electricity and magnetism and proved the existence of electromagnetic waves traveling at the speed of light. Maxwell suggested that light was nothing more than an electrical and magnetic phenomenon---electromagnetic wave---and were the same as Thomas Young's light waves.

This was another effort to explain matter. Just as mass was a characteristic of matter associated with gravitational force, electric charge was also a characteristic of matter associated with electromagnetic force.

Light then was an electromagnetic wave, which had a certain wavelength defined **as** the distance from crest to crest on a wave. Maxwell's equations demonstrated that a moving charge is capable of producing a magnetic field and a stationary charge produces an electric field. Also, he proved that electric and magnetic fields accelerate charges similar to the way gravity accelerates masses. And when these charges are accelerated, they give off electromagnetic radiation.

Now it was possible to speculate as to the nature of electromagnetic radiation. Maxwell's equations suggested that light was a wave of oscillating electric and magnetic field components. As a light wave propagated through space, the electric and magnetic fields produced would oscillate alternately with each other. The electric and magnetic fields are perpendicular to each other and perpendicular to the direction of propagation of the wave. In other words, the electric and magnetic fields flip, or oscillate ninety degrees, one with the other. The amount of oscillations per second is the frequency of the wave and is in the order of many trillions per second. The wavelength is the distance the wave will travel as the electric and magnetic field make one complete oscillation. The relationship between the speed of light, the wavelength and the frequency is given by the formula: Speed of light = wave length multiplied by frequency.

Concurrent with the above discoveries about light, other information was slowly evolving. The few natural philosophers of the world continued to ponder such matters while the rest of humanity was content with the biblical pronouncement of God: "Let there be light."

By this point everything known in the world of physics was explained by Newtonian Mechanics and Maxwell's electrodynamics. And for those few things unexplained, physicists were certain that the explanation would rest upon these two great theoretical foundations of physics. But try as they might, the concept of light and the atom remained mysterious. Man remained incapable of explaining anything at this level. Physicists had to try and resort to indirect measures in an attempt to understand these two mysteries.

Galilei Galileo (1564-1642)
made it possible to measure the speed of light. He did this by his invention of the telescope that allowed him to identify the four moons of Jupiter. This identification as well as his denying that the earth was at the center of the universe resulted in his being labeled a blasphemer and the Catholic Church banned his writings for two centuries.

It fell to
Olaus Roemer, (1644-1710),
a Danish astronomer, to take up where Galileo had left off. Since Galileo, natural philosophers knew that all four moons were in the same plane as Jupiter and therefore could be eclipsed by the great planet as they rotated in orbit around it. This made it possible to time the eclipse in relation to local earth time. By making these measurements when the earth was closest and farthest away from Jupiter, made it possible to determine the speed of light.

Since the French astronomer,
Jean Dominique Cassini (1625-1712)
had established that the distance from earth to the sun, or its radius from the sun, was ninety three million miles, the diameter of earth's orbit about the sun was one hundred eighty six million miles. It was determined that when the earth was a full diameter farther away from Jupiter, it took 16.6 minutes longer for the eclipse to be seen on earth. Dividing the one hundred and eighty six million miles by 16.6 minutes, or 996 seconds, gave a calculation for the speed of light as one hundred eighty six thousand miles per second, a figure that heaped ridicule on Roemer by the scientific community. They ignored his work. But fifteen years after Roemer's death a British astronomer,

James Bradley (1697-1712)
established that Roemer's figure was right on the mark. All this clever science had taken place in the seventeenth century.

Christian Huygens (1629-1695)
a Dutch physicist and a contemporary of Isaac Newton, promoted the wave theory of light. He believed that light, like sound, was a wave that passed through the mysterious aether. The wavelength and frequency of a sound wave determined the pitch of sound. The wavelength and frequency of a light wave determined the color of light.

THE SPECTRUM: AN INDIRECT PEEK INSIDE THE ATOM

At this time, Isaac Newton embarked upon a serious study of light when he passed the light of the sun through a prism: a three-edged piece of glass. He darkened a room, but allowed a single beam of sunlight to enter through a hole in the shutter and then pass through the prism. What emerged was multicolored light resembling a rainbow. These colors had a distinct order and never varied: red, orange, yellow, green, blue, indigo, and violet. He labeled this phenomenon "spectrum" (Latin for ghost). It was clear: the white light of our senses, bent when passed through the prism. The amount of bending depended upon the color. What could account for the spectrum? What was light made of that would allow the occurrence of this phenomenon? Since Newton was in favor of the atomistic theory of matter, he postulated that light was a particle like an atom, except that particles of light had no mass.

Subsequently, centuries later, it was realized that different colored lights would result when various forms of matter were heated, and when different rarified gases in cathode ray tubes were subjected to electric currents.

Georgius Agricola (1494-1555)
a pioneer in the field of mining, reasoned that it would be possible to learn the exact nature of the

burning material by noting the color of the spectrum. Dropping salt onto a flame would cause the flame to turn yellow. He discovered an interesting phenomenon, but practical applications would have to wait several centuries.

The next breakthrough, two centuries later, came when...

Joseph von Fraunhofer (1767-1826)
a German optician, managed to manufacture prisms of superb quality and produced spectra that were much more distinct then any previously produced. Looking at the sun, he observed many dark lines in the spectrum—Fraunhofer lines, but their significance was uncertain.

Then two professors from Heidelberg University provided some answers. One was a physicist,

Gustav Kirchoff, (1824-1887)
and the other was a chemist,
Robert Bunsen (1811-1899)
of Bunsen burner fame. They passed the light from hot gases through a slit and then through a prism in a box with a telescope: the forerunner of the modern spectroscope. The light was both magnified by lenses in the box and passed through the prism, thus making it possible for a greater detailed study of the light spectra emanating from various heated bodies of matter and gas.

What appeared were single colored lines known as line spectra. These lines depended on the gas, or other material tested, and were different from the characteristic colors of sunlight. Moreover, they discovered that each gas would emit its own characteristic color or wavelength. They discovered an excellent way to identify elements, and whereas chemists needed large quantities of these substances to make an identification, the burning of an element and its identification via line spectra was accomplished with as little as a fraction of a gram. Furthermore, mistaken analysis was not possible through this method.

This conclusion was crucial for further scientific inquiry into the nature of matter and, as it would turn out, the nature of the atomic structure of matter. If they burned any substance to a state of intense heat, the substance will emit light waves of a variety of lengths that result in a spectrum that is unique and different. As examples: potassium gives off a strong red line, sodium gives off two yellow lines, and the gas hydrogen gives off four lines: two violet, one red, and one blue.

If they cooled the element, it would absorb its own characteristic wavelength. White light or sunlight passed through the cooled element (gas) and the cooled gas absorbed the same wavelength of light that it would emit when heated.

The suggestion here was that there are certain energy states that are characteristic for each gas that are able to either give off or absorb energy. This fact would have great relevance in the future.

In the meantime, this development made possible the discovery of many new elements enabling the study of light from any source. This included the sun, other stars, and distant galaxies. In fact, they discovered Helium on the sun before its discovery on earth. Spectral analysis allowed physicists and chemists to study light in great detail. But, it was a study of a phenomenon that light could generate. It did not answer the question of exactly what light was and how it was generated.

As mentioned before, Sir Isaac Newton, believed that light was a corpuscular phenomenon. This atomistic theory postulated that light was made up of infinitesimally tiny particles of matter. His theory would prevail only to be replaced by Young's wave theory after the ingenious slit experiments.

Newton would have to wait 250 years for vindication.

ANOTHER PEEK INSIDE THE ATOM---THE ELECTRON---WHAT DOES AN ATOM LOOK LIKE?

Joseph John Thomson (1826-1911)
of Scottish descent, finally solved the mystery of the cathode ray. In 1895, in a set of two experiments using a cathode ray tube of his own design, he was able to confirm that the cathode ray had a negative charge. Secondly, it was not possible to separate the negative charge from the cathode ray. They were one and the same. This allowed him to state that cathode rays were charges of negative electricity. He reached this conclusion with a great degree of certainty, but the answer to the mystery of whether these negatively charged particles were the actual atoms, or molecules, or a particle of atomic matter was yet to be determined.

In a third experiment, Thomson was able to shift the cathode ray with an electrical field, and then forced it back to its original position with a strong magnetic field. Measuring instruments were able to determine the precise amount of the current and voltage necessary to accomplish this. From these measurements, he was able to calculate the speed of the negative particle which varied from twelve thousand to fourteen thousand miles per second depending on the voltage applied. The next step would be to calculate the exact mass and electric charge of the particle, but he could not make these discrete measurements. What was possible, though, was a calculation of the ratio of mass to electric charge. When he made this calculation the results showed that the ratio was almost nineteen hundred times smaller than a charged hydrogen atom.

He defined the cathode ray. It was made of infinitesimally tiny matter, with a negative electric charge, which originated at the cathode. The further assumption could be made that an electric charge was the sum total of all these rapidly moving particles. They could be considered to be "atoms of electricity."

In essence, Thomson had discovered that these particles were the same regardless of the nature of the gas placed within the cathode ray tube, and these particles were lighter than the lightest matter then known. The conclusion from these two facts alone was that the particles had to be a basic constituent of all matter, and if they were merely a part then there must be a whole. The atom was finally justified by experiment.

The Irish physicist,
G. Johnstone Stoney (1826-1911)
named these particles "electrons."

Electricity, then, was the free flow of electrons moving rapidly through space and not bound to any atom. Thomson opened the door to modern physics.

Since the size of an atom was known, and since the weight of the electron was determined to be approximately 1/1900th of the weight of an atom, and since the atom was electrically neutral, there had to be a positively charged component in the atom in order to balance the electron's negative charge.

Physicists now had to make sense out of the fact that a very tiny, light electron with a negative charge plus a positive charge that contained the remaining 1,899/1900th of the weight equaled the complete atom.

Thomson now proposed an atomic model that included a positively charged body in which was imbedded electrons, enough to balance the positive charge, and scattered in position away from each other. They could not be in proximity to each other because like charges repel. Therefore, the electrons had to be far enough away from each other to be at rest and in equilibrium with themselves and the positively charged substance. He likened it to seeds in a watermelon.

This atomic model became an accepted hypothesis, but lacked proof.

A BETTER VIEW OF THE ATOM

It would take a pupil of Thomson

Ernest Rutherford (1871-1937)
of New Zealand, to experimentally overthrow the Thomson model. Rutherford, a descendant of a Scottish family that had moved to New Zealand, was a brilliant student who took an interest in physics at the age of ten. He received a small scholarship and attended both Nelson College and Canterbury College. When he was twenty-three, he applied for a scholarship to Cambridge University in England. He came in second, but the first place winner declined, thus permitting Rutherford to get the scholarship. The world of physics would be intensely indebted to whomever it was that refused that scholarship.

Rutherford was working on x-rays when the discovery of radioactivity by Becquerel and the Curies turned his attention to this subject matter. He began work on the radioactivity produced by uranium. Much research done in this field led to the conclusion that the atom was not as stable as had been conceived by scientists going back as far as the ancient Greek philosophers. Evidence now suggested that radioactivity was a property of certain elements emitting subatomic particles and energy spontaneously. In so doing, the element would become something else.

Rutherford concluded that there were two types of radiation produced from uranium: the beta particle that proved to be an electron; and the alpha particle, which remained a mystery. A third type of radiation, similar to x-rays in its characteristics had a much shorter wavelength. This was the gamma particle or gamma ray.

In 1902, Rutherford put forth the theory that radioactivity found in some substances was nothing more than spontaneous decay accompanied by the emission of differing rays of an electromagnetic and material nature. This struck the scientific world like a bombshell and overturned the concept of an indivisible atom of final and unchangeable structure. In essence, a certain group of elements could spontaneously decay by the release of the beta and alpha particle and gamma rays, and become another element lighter than the parent.

Rutherford moved to McGill University in Montreal where he discovered the nature of the alpha particle. It was positively charged and identical to a helium atom containing two protons and two electrons, and it accelerated with a great deal of energy.

He set out to prove that Thomson's watermelon view of the atom was correct. He struck gold foil with the alpha particles, and since they were seven thousand times heavier than the electron he ignored the influence of the gold atom's embedded electrons. The positively charged portion of the atom, if struck with alpha particles, should likewise not offer any resistance to the alpha particle that would therefore pass right through. But, to Rutherford's amazement, one in twenty thousand did not. They were reflected right back. Although the figure of one in twenty thousand seems very inconsequential, when one considers the billions of alpha particles projected through the gold foil the number is far beyond the virtually nil probability of such an occurrence. To quote Rutherford, "It was quite the most incredible event that had ever happened to me in my life. It was almost as incredible as if you fired a fifteen-inch shell at a piece of tissue paper and it came back and hit you."

He reached only one conclusion: the Thomson model of the atom was incorrect. In order to account for the alpha particle deflection, there had to be a concentrated bulk of positively charged particles occupying the center of the atom where almost the entire mass of the atom was located. When the alpha particle collided with this mass, it deflected back. Rutherford named this obstacle to the alpha particle the atomic nucleus.

Rutherford's model of the atom included a positively charged atomic nucleus containing almost all the mass. Electrons, with their negative charge revolved around the nucleus. The configuration resembled a miniature solar system including a great amount of space between the nucleus and the electrons similar to the space between the earth and the sun in relative terms.

If this were true, then what kept the electrons from not spiraling into the oppositely charged nucleus? The proposed answer at the time was that the attractive or centripetal force of the positively charged nucleus was counterbalanced by the centrifugal force of the negatively charged electrons as they revolved around the nucleus with incredible speed.

Rutherford's model was compatible with current classical physics theory of Newtonian Mechanics and Maxwell's Electrodynamics except for one fact: Maxwell's electromagnetic theory proved that when electrons accelerate, light is emitted. If this is the case, then Rutherford's revolving electrons should give off light. But, when they give off light they lose energy, and if they lose energy then there is a reduction in the electrons centrifugal force. If that is true, the electron's centrifugal force lessens and the attractive centripetal force of the positively charged nucleus would be greater, and if it became greater why wouldn't the electrons spiral into the nucleus? This means all atoms would collapse---and in millionths of a second. Fortunately, that is not the case. Most atoms last forever. More work was necessary.

OPEN THE DOOR TO A NEW PHYSICS

Max Planck (1858-1947)
a German physicist, would stun the physics world. He was an authority on thermodynamics and wrestled with the problem of blackbody radiation.

He performed heat studies on a closed metallic black box within which was a vacuum and through which there was a small opening. The box, heated, made the interior of the box the same temperature as the box itself. Clearly, heat then is transmitted in a vacuum, but this is no surprise because the heat of the sun reaches the earth after traveling ninety three million miles through the vacuum of space.

How is that heat transmitted? The answer: by light.

As the box is heated, the vacuum inside the box fills with light. And just as the heated metallic box goes from red-hot to white-hot, the light in the interior of the box changes colors as does the box itself. The color of the light in the box depends upon the temperature of the box. And the color of light depends upon its wavelength or crest to crest distance. A long wave length, or low frequency, and a short wave length, or high frequency, determines the color. Theoretically, the colors are infinite, but our eyes allow us to see only the colors of the rainbow. Low frequency light of four hundred trillion cycles per second---400,000,000,000,000 is red. High frequency light of seven hundred fifty trillion cycles per second---750,000,000,000,000 is purple. These and five other colors in between make up the visible spectrum of light: Red orange yellow green blue indigo violet (Roy G. Biv). Light that we cannot see includes radio and infrared waves of low frequency, and ultraviolet rays, x-rays, and gamma rays of high frequency. They are all part of what we know as the electromagnetic spectrum.

Fortunately for us, even though we live surrounded by many forms of light making up the entire electromagnetic spectrum, our eyes can only sense the tiny fraction between red and violet. Thus the world's beautiful colors are open to our senses.

Physicists like Max Planck were intensely interested in identifying the spectrum of light within the vacuum of the black box. One could not see inside the atom, but the key to unlocking the atom's secrets lay in the spectrum, because every atom of each of the elements produced a fixed spectrum. Since the atom was composed of a nucleus and electrons, and since it is the electrons that move around a stationary nucleus, it is the electrons that produce the spectrum. Therefore, the study of the spectrum will allow you to learn about the movements of the electrons, thus helping to clarify atomic structure and motion.

Planck wanted to understand the nature of the light that fills the box as it is heated and exits through the small hole in the box. The color of the light within the box varies with the temperature and this he plotted using a line graph measuring the frequency on the horizontal axis against the intensity, or brightness of the light, on the vertical axis. The frequency of light with the highest intensity determined the color within the box, and since the frequency is infinitely variable, the line was thought to be continuous. This is an important thought and Planck unknowingly was about to make an assault on this concept.

The physicists wanted to understand why a particular light spectrum existed at any temperature, and the first efforts to explain this were two Englishmen. They used classical, accepted physics theory (Newton and Maxwell) in their work.

John William Rayleigh (1842-1919) and James Jean (1877-1946)

were both English physicists who developed a formula to describe the box's spectra. They assumed that every wavelength would be radiated in the black box's radiation, and they felt confident their formula would predict the exact spectra. But, it did so only for low frequency light. This exploded like a bombshell on the physics world. Classical theory of Newton and Maxwell could not be wrong!

Wilhelm Carl Wien (1864-1928)

developed another equation that he hoped would correct the problem, but it did not. His formula matched experimental results at high frequencies, but it was a little off at the lower ones.

With the work of these two physicists, the world of classical physics, and Maxwell's electrodynamics was under assault.

Max Planck decided that the two formulas should be merged in some way in order to match the experimental results. After four years of intense work, he succeeded. He developed a formula that transformed into Rayleigh-Jeans at low frequencies, and transformed into Wien's at high frequencies. Planck's equation explained black box radiation perfectly. He took the constants in the equations and merged them into one. This would become known as Planck's constant. (h). It was an infinitesimally tiny number (.00000000000000000000000000662...) 6.62 octillionth, or 6.62 billion billion billionth.

And his constant and his formula changed physics forever, because the only way his equation could fit was if the light energy had discrete values as opposed to the classical accepted theory of the time that stated that light wave energy and frequency varied continuously. Therefore, the idea of discrete values was unthinkable, and physicists, including Planck himself, did not accept his own theory, which was that light energy is not continuous. It peaks in a discontinuous manner, and the discontinuity of Planck's formula is due to the fact that light is a particle, and its discrete energy values were unbelievably feeble.

Albert Einstein (1879-1955)

a German-Swiss physicist and mathematician, was unknown before 1905 when he was twenty-six years old and working in a Swiss patent office. He was young enough to dare to question the established order, and when he heard about Planck's discovery his reaction, as opposed to most in the physics world, was one of credibility and acceptance. He suggested that thinking of light as a particle was explanation enough. And the particle has energy (E) equal to Planck's constant (h) times the frequency of the light (v). $E=hv$. We now see where Planck's constant fits.

The most famous constant of all is Pi (Π) (3.14159…). Like $E=hv$, we learned in grade school that $C=\Pi D$, or the circumference of a circle equals Π (a constant) times diameter. In other words, if we know the diameter of a circle, then by simply multiplying it by Π, a constant stable number, we will always immediately know the circumference for any circle no matter how big or small. Also if we know the circumference, we can immediately learn the diameter by dividing the circumference by Π. Also, every time you divide any circle's circumference with its diameter you will always get Π. This is the nature of a constant, and constants have a crucial place in mathematics and physics.

The same is true for $E=hv$. By knowing v, the frequency of light, we would always know the energy by the simple expedient of multiplying the frequency by Planck's constant, h. And by knowing the frequency of any light and its energy we can experimentally find Planck's constant h.

In 1905, Einstein published three papers in the *Annalen der Physik.* Never before or since has there been such an amazing intellectual output by one man in one journal. The first was the light quantum or photoelectric effect, the second was his

Brownian motion paper that proved that atoms exist, and the third was his famous work on special relativity.

In the first paper, he concluded that light consists of energy quanta known as photons, which can penetrate a metallic body and transfer part of their energy to the electrons resulting in their ejection from the metal. Before an electron ejects, it has to perform an amount of work necessary to get out of the metal and into the air. Those electrons closest to the surface of the metal will have the least work to get free. Einstein described this with simple equations. He was thus able to explain that the number of electrons ejected did not depend on the frequency of light, but the velocity of the electrons did. He did show that the electron's velocity was indeed directly proportional to the frequency. He used Planck's work to explain, stating that the energy of the light transferred to the electron in discrete amounts (photons) each of magnitude hv where h is Planck's constant and v is the frequency of the light.

We can demonstrate this with the high frequency ultraviolet rays of the sun. The higher the frequency the higher the energy (E=hv). When these high-energy ultraviolet rays strike our skin, the electrons are knocked off with great force and cause a reaction (sunburn). Low frequency infrared rays with less energy, emanating from a heating source such as a stove, do not cause an intense reaction and therefore we learn that a stove is a poor way to get a tan.

Einstein's work was confirmed by an American physicist...

Robert A. Milliken (1868-1953).
In the second paper, Einstein was able to prove that minute particles like pollen grains suspended in water exhibited unceasing and irregular random motion due to the fact that the molecules of water were bombarding them. This was further proof of the atomic theory of matter.

The third paper, Einstein's special theory of relativity, made it possible to predict experimental observations anywhere from a velocity of zero up to and including light speed, one hundred and eighty-six thousand miles per second or three hundred thousand kilometers per second. What this meant was that the 200 year old Newtonian mechanics, now became a specialized case of Einstein's theory.

Einstein's special relativity is based on two main postulates. The first is: *the laws of physics remain intact in all inertial reference systems.* Put another way, the laws of physics are valid for all frames of reference in uniform motion (we're back to Galilean relativity). Although Galilean relativity referred to the laws of motion, Einstein expanded it to also include electromagnetism---traveling at the speed of light.

To understand this concept better, let's say that you are performing a physics experiment here on earth. You are in a frame of reference, or inertial reference system, that is at rest with the laboratory since you're sitting at your lab desk. I drive by at a fixed velocity with respect to your laboratory and I can watch you through an open window in your lab. I am also in an inertial reference system, but mine is moving at uniform motion with respect to your lab. Since you have found the laws of mechanics to be true in your lab, I must also agree with your findings. And this is valid in accordance with Newton's first law of motion that states that the laws of mechanics are the same for all inertial reference systems. Put another way there is an equivalence of the laws of motion in different inertial frames.

A second illustration is as follows: You are standing in an airplane aisle while the airplane is traveling at 550 miles per hour. You throw a ball in the air and notice that it comes right back down to your outstretched hands. You have observed the ball traveling in a vertical path and you note that the ball has followed the laws of motion and gravity and acted exactly as it would have acted had you been standing on earth.

I am standing on the earth and watch you fly by. I can see into the plane and watch as you throw the ball in the air. As a stationary observer at rest on the surface of the earth, I observe the ball tracing out a parabola and traveling the same speed as the airplane in the direction of the planes travel. So we have seen the same event differently, but we both agree that the motion of the ball follows the laws of motion and gravity. Or, put in proper physics language: in describing the laws of motion there is no preferred frame of reference.

So far, we have dealt with relativity, but actually a form of relativity brought forth by Galilei and Newton. So what has Einstein to do with this development? Well, he confirmed their concept and added a unifying principle that was another giant leap forward. For he answered another question: in what frames of reference is Maxwell valid? This was a crucial question because put another way, in what frame of reference does light travel at speed C or one hundred and eighty-six thousand miles per second? And this brings us to Einstein's second postulate. It states that *the speed of light is the same for all observers regardless of their motion or the motion of the light source.*

To understand this better let us go back to the principle espoused by physicists up to the end of the nineteenth century. And that is that, like water waves requiring a medium of water in which to travel, and sound requiring the medium of air, light too, as a wave, required a medium. And thus, the aether was postulated to serve as the medium for light waves to travel within.

Physicists postulated that the aether had to permeate all space and had to be tenuous. But at the same time it was tenuous it had to be stiff in order to propogate a wave traveling at such enormous velocity. Since there was an aether permeating all space this meant that the earth was traveling through it as well. Or was it? If it wasn't, then the earth as opposed to all the other planets and stars were uniquely at rest and this was already demonstrated not to be the case when the heliocentric view of the cosmos was proposed by Copernicus. Another possibility was that the earth drags the aether along with it as it travels through space. This was proven not to be true by a phenomenon of starlight aberration. The last suggestion was that the aether was blowing past the earth. And if this was true then the speed of light waves should move faster when measured traveling in the direction of the aether wind: one hundred eighty six thousand miles per second plus the speed of the wind (the speed of the earth or 20 miles per second) or one hundred eighty six thousand and twenty miles per second. Conversely when traveling against the aether wind the speed of light should be one hundred eighty five thousand nine hundred and eighty miles per second. Neither of these was true.

The implication then was that the earth is not moving relative to the aether and the speed of light is the same in all directions. This was a major contradiction that struck physics. Rather than abandon the cherished aether concept another explanation was sought.

Two physicists
Henrich Antoon Lorentz (1853-1928) a Dutch physicist, and...
George Francis Fitzgerald (1851-1901) an Irish physicist
independently developed a theory stating that a body moving through the aether with a certain velocity contracts by a specific amount. And this

shortening would compensate for the slower speed of light that physicists still insisted would happen as the beam of light moved against the aether wind. This was their explanation that would allow the concept of the aether to remain intact.

This set the stage for Albert Einstein who boldly stated that the concept of the aether was a fiction that should be done away with and he presented his theory of special relativity that was another unifying leap forward. For now, not only the laws of motion were the same for all observers in uniform motion, but in addition, this concept held true for the laws of electromagnetism as well. He also stated that the speed of light was a constant and did not depend upon the speed of its source.

What Galilei and Newton knew held for the laws of motion was true also for electromagnetism.

Newtonian Mechanics works perfectly for our everyday worldly experience, but at speeds approaching the speed of light, such as on an atomic scale, Einstein's relativistic concept must be utilized.

What are the consequences of Einstein's theory of special relativity? They are unusual effects that seem to defy common sense.

First: Clocks in motion slow down. This is known as time dilation and its effects are principally at speeds very close to the speed of light. Taken to its extreme; at the speed of light time stops.

Second: The length of any object in motion is contracted in the direction of that motion.

Third: Events that appear simultaneous in one frame of reference may not be simultaneous in another frame of reference moving in relation to the first. Which observer is right? The answer is that they both are the—relativity of simultaneity.

Four: The mass of a body increases as its speed increases. The effects of this are noted principally at speeds approaching the speed of light. Taken to its extreme, a mass could never travel at light speed, because the energy to accomplish this task becomes infinite if light speed is reached—therefore not possible. Saying it another way: any mass traveling at light speed becomes infinite. This means it would weigh more than the universe. Again, clearly not possible.

In ten years, Einstein would add his general relativity theory to his accomplishments. We will discuss this later.

By 1885, physicists firmly established that each atom produced light of a specific frequency. Although physicists were able to determine the kind of light produced by many different atoms, they did not understand how only certain frequencies were generated.

According to Maxwell's theory, Rutherford's atomic model should glow with light of infinite frequencies. However, atoms would glow only with frequencies characteristic of each atom. These frequencies do not change and were as important to identification of an element as DNA (in the modern era) is to the identification of a specific person.

Johan Jacob Balmer (1825-1898)
of Switzerland was a mathematics teacher who enjoyed seeking patterns in numbers. He became interested in the hydrogen spectrum and was able to demonstrate that the spectral lines followed a simple mathematical formula based on integers. This formula described the relationship of the four different wavelengths of the spectrum of hydrogen. For the first time physicist established that atoms emitted spectra in a regular pattern. And it was this type of regularity that fascinated Balmer and resulted in the formula that would add his name to the immortals of physics history. In addition, by the use of his formula, Balmer was able to predict more lines in the ultraviolet and infrared frequency range which, unknown at the time, would subsequently be discovered.

It became abundantly clear that the emission or absorption of light from an atom reflected the increase or decrease in its energy. In some manner, the atom was rearranging itself and this rearrangement was reflected in its emission and absorption spectra.

But this clarity was slow in coming and Balmer finally received accolades for his amazing formula, but not during his lifetime.

Johannes Rydberg (1854-1919)
discovered that Balmer's equation had relevance in finding frequency as well as wavelength. With his equation, it became possible to identify all the spectra given off by any atom, not just hydrogen.

As an example, physicists knew about the four visible spectra of hydrogen, but Rydberg's formula identified many more, all eventually confirmed experimentally, just as Balmer predicted.

Balmer, Rydberg, and others had taken a leap forward in understanding the spectra of atoms, but the mechanism of why these spectra occurred remained a mystery.

Then John Thomson discovered the electron and Ernest Rutherford clarified atomic structure.

Before we go on to Niels Bohr, mention should be made of…
Thomas Rees Wilson (1869-1959),
a Scottish physicist, who developed the cloud chamber in 1912, one year before Bohr's groundbreaking work.

The cloud chamber is an apparatus consisting of a glass tube filled with air and water vapor under much reduced pressure. When alpha and beta particles were propelled into the chamber, fine cloud tracks became visible. These tracks were made possible by the ionization of the air within the chamber. When the air is ionized it takes on an electric charge, and in this state, it has the ability to attract the water vapor forming a cloud. At first when the particle entered the cloud chamber, it caused many cloud droplets to form. This resulted in a photograph that demonstrated the large cloud rather than a single track outlining the particle. This problem was eventually corrected by a strong electric spark shot into the chamber, which lasted only long enough to light up the tracks, but made a photographic record possible. Eventually, physicists could take more than 100 photographs per minute. When the cloud chamber is used together with a strong magnet, it becomes possible to identify a particle by its degree and direction of deflection.

Wilson's clever invention allowed a clear fine track of alpha and beta particles as well as making it possible to discover other particles.

A NEW ATOM. THE START OF QUANTUM MECHANICS

Neils Bohr (1885-1962)
was a Danish physicist. His father was a famous physiologist and his mother was the daughter of a prominent Danish Jewish banker. This fact, plus his Jewish wife, would cause him to flee Denmark when Hitler ordered all Danish Jews rounded up during the Nazi occupation in World War II.

The story of how the courageous Danes saved the great majority of Danish Jews is inspirational, and the reader is encouraged to learn the details of this heroic effort.

In 1913 Niels Bohr revolutionized physics.

At age twenty-six years, Bohr's thoughts were on three relatively new and paradigm shifting basic physics facts.

* Rutherford's planetary model of the atom with negatively charged electrons rotating around a much larger positively charged nucleus balanced by centrifugal and centripetal force. The unanswered question remaining was why do the electrons not spiral into the nucleus when they give off energy in the form of light?
* Planck's energy 'quantum' theory, which stated that the energy of light has discrete values.
* Einstein's light 'quantum' theory that stated light is a particle with energy equal to hv.

A quantum is an elementary unit of energy. Quantum theory is a physics theory based upon the concept that radiant energy, such as light, is composed of small discrete packets of energy.

Bohr had no choice but to resolve these problems by dealing mathematically with spectra, for all physicists knew that in the spectral colors and lines were hidden the secrets of otherwise inaccessible atomic structure.

Bohr reasoned that since Planck's formula can have only specific discrete values, then the atom's energy must also have discrete values. If the atom's energy has discrete values, then the orbits of the electrons must have discrete values as well.

This brilliant thought experiment resulted in Bohr developing an atomic model consisting of a nucleus at the center with electron orbits each having its own discrete value. As long as the electrons keep moving in their orbits, no energy is used and the atom is considered to be in a stable state. If energy is not used in this stable state, then the electrons will not spiral into the nucleus.

Bohr stated that if a unit (or quantum) of electromagnetic radiation is absorbed by the electron in an atom, the electron energy will increase and it will jump to a higher, outer orbit with greater energy then the inner orbits. For an electron to go from a higher orbit to a lower orbit (one with less energy level), it will have to release a quantum of electromagnetic radiation. When it releases this quantum, a line spectrum with a certain frequency appears spectroscopically.

The essence of Bohr's theory is that when electrons jump from one orbit to another they emit light. The light particle emitted, a photon defined as a quantum of electromagnetic radiation gave birth to quantum theory.

Bohr had taken a bold and provocative step. In one powerful sweep he dared to contradict Maxwell's electrodynamics which would cause an electron moving around a nucleus to radiate energy and spiral into the nucleus. He simply stated that electrons revolving around a nucleus will not spiral into the nucleus. And the reason is because the electron is rotating in a stable orbit where its energy is fixed. It does not emit or absorb energy.

If one could suddenly see an atom expand to the size of an average room, the electrons would be revolving in an elliptical orbit while the nucleus was the size of a speck of dust. Atoms are mostly empty space.

We know that the electrons emit light, but how do they do it? The answer brings us back to Einstein's hypothesis about photons. Remember $E=h\nu$. Each light unit has energy equal to $h\nu$ (planck's constant times frequency). And, as has been stated, this energy has discrete values.

Bohr stated that when an electron is in the outermost orbits it has greater energy, and when it "jumps" to an inner orbit it does so by liberating a photon with its energy equal to $h\nu$. Conversely, were an electron to absorb a photon with energy $h\nu$, it would become more energized, and jump to an outer orbit more capable of handling an energized electron.

There is a basic principle of physics called the law of conservation of energy. In essence, the law states that in any system the energy remains constant. Did Bohr violate this law by his jumping electrons that gain energy when they jump from an inner to an outer orbit, and lose energy when they jump from an outer to an inner orbit? The answer is no, and Bohr could turn to Einstein and utilize his photon to keep the law intact. For when an electron loses energy in a jump, the energy lost emits in the form of a photon of light with a certain characteristic frequency. And when an electron absorbs energy, it does so when it grabs onto a photon of the correct frequency.

This energy, calculated by Planck's famous formula: $E = h\nu$ means that when the electron absorbs a photon, its total energy is now equal to the separate energy of the photon plus the energy of the electron before the absorption. Conversely, when an electron liberates a photon of a specific frequency or color, the electron's energy becomes less by the amount of the liberated photon's energy.

Either way the law of conservation of energy remains intact.

These theories allowed Bohr to develop an equation expressing the frequency of an atom. He noticed that his new equation and Rydberg's equation describing the spectra of hydrogen had the same form. This meant that if these two equations were equivalent his new theory would be correct. By the use of some clever basic mathematical manipulations Bohr established their equivalency and declared that energy levels become greater the farther away the orbit is from the nucleus, but as the energy levels increase in the outer orbits, the difference in energy between these distant, adjacent orbits decrease. Remember this fact, which will soon have great relevance.

As Bohr predicted, energy levels exist in a discontinuous stepped form.

An interesting sidelight to Bohr's work was that for the first time, by using an equation developed by Bohr, it would become possible to calculate the radius of a hydrogen atom. For hydrogen it proved to be 0.000000053 centimeters or fifty-three billionths of a centimeter). Since the diameter of a circle is twice the radius of the atom's length, the diameter is 0.000000106 centimeters or one hundred and six billionths of a centimeter.

Physicists had long been accustomed to working out the mechanics of large bodies: the earth and the planets, cannonballs, bullets, freight trains, etc. Newtonian classical mechanics worked perfectly at this level. Small wonder that they had such difficulty and difference of opinion when it came to bodies the size of the atom where benefit of sight was lacking and all physicists had was indirect evidence of atomic machinations such as spectral lines.

However, it had yet to be proven that there actually were stationary states with discrete energy values within the atoms themselves. Theoretical ingenuity and mathematical wizardry were impressive enough, but experimental proof was lacking, and a theory was only a theory until proven experimentally.

In 1914, one year after Bohr's hypothesis,

James Franck (1882-1964) and
Gustav Ludwig Hertz (1892-1985)
a nephew of Heinrich Hertz, bombarded mercury atoms with electrons and found that the mercury atoms absorbed energy from the electrons, but only over a certain critical value. They assumed that they had ionized, or given a charge to the outermost electrons in the mercury. But when Bohr heard of their experiment, he disagreed with their interpretation and suggested that the energy change was the result of the absorption of quanta. Frank and Hertz stuck to their theory for two years until they finally realized that Bohr was right. Unknowingly then, Frank and Hertz experimentally confirmed that there were stationary states with discrete energy values. Bohr's theory now had a firmer foundation.

Bohr had truly revolutionized atomic structure and function. No longer was there any question that the use of quantum theory could explain the nature of matter.

This abrupt leap into a new physics had some physicists concerned. Was classical physics being overturned or was this theory consistent with classical theory?

We have already stated that when an electron rotates in distant orbits, the energy level between these orbits virtually disappears. Also in classical theory, electrons liberate light as they rotate. In Bohr's theory, electrons liberate light as they transition from one orbit to the next. But in the outer orbits where the energy levels virtually disappear, the energy levels for the hydrogen atom using Bohr's theory and classical theory are the same. He fit his quantum theory into classical theory rather than abandon classical theory and replace it by quantum theory. This was called the Correspondence Principle and would satisfy many of his critics, but not all.

In one year Bohr shattered Maxwell's electrodynamics, developed the stable orbital theory of the atom, and did not violate the law of conservation of energy when he allowed the energy differences to be explained by Einstein's photon hypothesis.

It makes one believe in reincarnation when Bohr's birth date (1885) was the year Johan Jacob Balmer died.

Bohr's correspondence principle initially served to quell some critics, but in time it became evident that some of Bohr's theoretical analysis could not be matched by experimental data.

A NEW GRAVITY

For any observer in a state of uniform motion, the same laws of physics hold. This was Einstein's special theory of relativity, and this intellectual leap forward revolutionized physics.

Approximately ten years later, he announced his general theory of relativity. This further unified physics when the theory added an area where physics laws would also hold; and that was non-uniform motion (acceleration). Therefore, this new theory removed the special relativity restriction to uniform motion and included non-uniform motion as well.

The principles of special relativity were firmly established, but Einstein felt he had to reconcile gravity with these principles for there now were clearly defined inconsistencies between Newton's theory of universal gravitation and the principles of special relativity.

First: Newton's law of universal gravitation defined gravity as a force acting instantaneously through a distance. The earth affecting the moon, and the sun affecting the earth, for instance, had to be instantaneous effects. But, this concept was incompatible with special relativity's self-imposed speed limit: the speed of light. For how could there be instantaneous forces if nothing could go faster than this limit?

Second, according to Newton's theory of gravity, the force that bodies exert on each other was inversely proportional to the square of the distance between them. This meant that for every one unit of distance apart, the force would decrease by the square of the distance. At two units apart the force would be one fourth, at three units apart the force would be one ninth, at four units apart the force would be one sixteenth and so on. But special relativity showed that distance and simultaneity are relative and so these different results in different reference frames could not be compatible with special relativity. Clearly, a unifying principle would have to depend upon a different concept of gravity.

According to Newton's second law (F=MA) force equals mass times acceleration. It takes a greater force to accelerate a larger mass object then a smaller one. This mass, termed inertial mass, is the property of a body to persist in a state of rest or straight-line uniform motion. This state will only change when acted upon by a force. The force of gravity depends upon an object's mass as well, and this is termed gravitational mass. Inertial and gravitational mass are the same. From this relationship, it is clear that gravity and acceleration are related. This principle is termed the equivalence of gravitational mass and inertial mass, or the principle of equivalence. Stated another way: the effects of gravitation and the effects of acceleration are indistinguishable one from the other

The example of the falling elevator demonstrates this. If one is unfortunate enough to be in an elevator whose cable sudden breaks, the elevator falls at the usual accelerating rate of thirty-two feet per second per second. Anyone in the elevator will be floating and if one removes a coin from his or her pocket and drops it, it will float alongside only to cease tragically of course, the moment the elevator hits the floor. For a brief time you were in a state of free fall, and this is similar to the free fall of a satellite which is orbiting the earth. When you are in free fall in the plummeting elevator, and the astronaut is in free fall around the earth, neither of you will feel gravity, and you will float within your confines.

If your spaceship is on its launch pad on the surface of the earth and one drops a coin it will fall to the floor of the spaceship. If the spaceship is traveling in outer space and one drops a coin at the same instance he or she accelerate the ship, the floor of the space ship will come up and hit the coin, and it will be indistinguishable from the coin falling to the floor. This demonstrates that acceleration and gravity have similar effects. At least this is true for the typical small gravitational effects consistent with our world.

Through all these thought processes Einstein formulated his own theory of gravity. He came to the following conclusion: the general theory states that all laws of physics should be the same in all reference frames. Since, in the freely falling reference frame of the elevator or the orbiting spaceship one does not feel gravity, we have somehow, in these reference frames, been taken away from gravity's effects. Therefore, gravity cannot be real. If it is not real and it is not a force obeying all physics laws in all reference frames, what is it? When Einstein gave his answer he made what most people consider one of the few greatest intellectual achievements to come out of the mind of man in all of human history. His name could now be mentioned in the same sentence as Galillei, Newton, and Maxwell.

For Einstein said gravity is not a force. Rather it is geometry of space and time. It has geometric properties, but does not follow the laws of Euclidean geometry that we all learned in high school. Rather, space has four dimensions, the fourth being time. And any object moves in the straightest possible path through curved space-time, and it does so because it is affected by the neighborhood within which it finds itself and its movement is dependent on the mass of nearby objects that are curving (warping) local space-time. The greater the mass of the nearby object the greater will be its influence and the greater will be the curvature of the body's path moving through space as the body seeks its straight line path.

This ended Newton's force of gravity manifested as a force exerted between distant objects. Gravity was not a force at all but instead it took on Einstein's new view of the universe. What causes space-time to curve? Einstein said matter and energy; and from this came many predictions. The next seventy years would verify them all.

In our solar system where gravity is relatively weak, Einstein's predictions, based upon general relativity, differ insignificantly from Newton's theory, but the stronger space-time is curved, the greater can be the effects; the so called relativistic effects. This gives rise to the following predictions:

First: elliptical orbits do not remain fixed in space but rather rotate slowly about the gravitating body (procession effect). The more massive the gravitating body the more pronounced the procession. In fact, in 1919 physicists proved this concept for the planet Mercury.

Second: the greater the space-time curvature the slower time will run. This is attributed to the fact that light, which can't slow down, loses energy trying to escape from a gravitating mass. And when it loses energy its frequency is reduced and to an outside observer time will appear to be slower (gravitational time dilation).

Third: light, bent by curved space-time is proven by many astronomical observations.

Fourth: We have learned that the greater the mass the greater the warping of curved space-time. Taken to an extreme the theory postulates black holes, so named because gravity (curved space-time) is so massive around one of these bodies (collapsed star) that even light emanating from them cannot escape because of the powerfully massive gravitational effects.

Einstein became the most famous physicist in the world and his name became synonymous with the word genius.

A few final words on his famous formula $E=MC^2$. When Einstein derived this famous relationship, he never did so with a nuclear (atomic) bomb in mind. Contrary to what most people think this formula is not the formula for the nuclear bomb. Rather Einstein made the argument that inertia (a property of matter by which it remains at rest or at uniform straight-line motion unless acted upon by an external force) has energy. The more external force that exerts upon an inertial mass, the greater will be the acceleration. But this acceleration can never reach the speed of light because at that speed the mass (that has been getting heavier and heavier as the force acts upon it) will become infinite. Clearly, that is an impossible result. So Einstein never talked about being able to get all that energy out of any mass. It took others, many years later, to understand how to release all that pent up energy and change the world forever.

The statement made previously that uniform straight-line motion represents the order of things in our universe is counterintuitive to us as we stand stationary on the surface of the earth. When we stand, it feels like we are at rest. We are not moving. But we have learned that there is no such normal state as rest. The earth rotates on its axis, and while doing so it revolves around the sun. The sun and the entire solar system revolve around the galaxy, and the entire galaxy moves in our expanding universe.

The idea of standing still at rest is impossible in a world where everything moves, and does so at incredible speeds. There is no resting state in such a universe. Just think of one of the satellites that

NASA has rocketed into space to travel through and eventually leave the solar system. Even though it has used up its rocket fuel, it nevertheless continues in rapid straight line uniform motion, and will persist in this natural state until some intelligent being from another galaxy stops it, or it travels too close to a large body whose gravity, or space time warping, will be the external force that changes the natural straight line path.

We have the benefit now of knowing about all these motions on our earth; in our solar system; in our galaxy; and through our universe. The natural philosophers who decided centuries ago that uniform straight-line motion was the normal state had no such benefit of all these motions and their wisdom is therefore all the more amazing.

BACK TO THE ATOM

Bohr took his work to
Arnold Sommerfeld (1868-1951)
a brilliant German mathematician. Bohr's formula used whole numbers in one of its terms. This was because whole numbers of energy pockets or quanta were involved. The theory postulated this need. It was not possible to have fractions of discrete energy. Any whole number used in Bohr's formula to replace the specific term in his equation was therefore called a quantum number.

With increasingly sophisticated spectroscopic instruments, it became evident that some spectral lines were actually made up of numerous thinner lines. Could the Bohr orbits actually be a number of orbits with small differences? How could these spectroscopic findings be explained?

Using mathematical techniques as only he could, Sommerfeld proved that the electrons not only moved in a circle, but also rotated about the nucleus in elliptical orbits. He also showed that the electron need not move in the same plane. Sommerfeld's improved Bohr atomic model utilizing the quantum number of Bohr plus his own explained the spectrum of hydrogen much better.

Pieter Zeeman (1865-1943)
a Dutch physicist demonstrated that different spectral lines appeared when an atom was exposed to a magnetic field. This work was done in the early

1890's and preceded Bohr's quantum theory. This occurrence became known as the Zeeman Effect. Any future atomic theory would have to take this effect into account. Clearly, exposing an atom to magnetic influence energized the electron and caused it to select another orbit pointing in a direction dictated by the field. Sommerfeld now tackled this problem and solved it as well.

He added a magnetic quantum number. There were now three.

Bohr knew that his theory was incomplete and like all theories would and should be subjected to the kind of scientific scrutiny so necessary for advances to take place. He welcomed them and congratulated Sommerfeld on his brilliant work amplifying and clarifying Bohr's theory. Bohr was cooperative with all the physicists working on the quantum theory, and his institute in Copenhagen became one of the major collecting places for all new concepts to be studied and subjected to intense scrutiny. The institute became world famous.

Einstein also praised Sommerfeld for his insight. But, had they completed the work? Not quite. Something was lacking. Further magnetic results demonstrated that there were more unexplained spectral lines that no one could explain. This mystery was given the name of the Anomalous Zeeman Effect (AZE). It would take a young Austrian physicist to sort this out in the mid 1920's.

Einstein's˙ photoelectric effect suggesting that light was a particle did not convince all physicists, but eighteen years after Einstein's hypothesis,

Arthur Holly Compton (1892-1962)

considered x-rays as composed of discrete pulses, or quanta, of electromagnetic energy. Since x-rays are a form of light (part of the electromagnetic spectrum), they have energy and momentum the same as material particles. They also have wave characteristics, such as wavelength and frequency. Compton scattered the x-ray photons through graphite and determined the change in frequency of the x-ray photon after it hit an electron. This enabled him to measure the direction and velocity of the electron. He used the concept of collisions like a billiard ball to predict the movement of the electrons. The cloud chamber confirmed his theory and demonstrated the electron paths. This work put the frosting on the cake of Planck and Einstein's photon hypothesis. Light exchanged energy as if it was a particle. There was no longer any doubt.

Wolfgang Pauli (1900-1958)

of Austria was a prodigy, as were so many of the physicists we have mentioned. At the age of eighteen, he started his training under Professor Arnold Sommerfeld at the University of Munich. He was twenty-one when he wrote a paper on Einstein's relativity theory. Einstein was amazed at the sophistication of Pauli's writing and called it "grandly conceived." This was more than enough to establish Pauli's reputation.

Pauli postulated that the anomalous Zeeman Effect was due to spinning electrons. This concept states that electrons, as they revolve around the nucleus, also spin similar to the way the earth rotates as it revolves around the sun. Two other physicists independently came up with the same theory. They suggested that the changed spectral line splitting of the anomalous Zeeman effect did not require a fourth quantum number, but was due to the spinning electron. They published their results and received credit for the development of the theory.

Pauli, however, was not to be undone. The spin of the electron could be clockwise (spin up) or counterclockwise (spin down), and he calculated a spin quantum number as $+1/2$ and $-1/2$, indicating that the electron could have two possible states. The energy of the electron is slightly different for these two spin directions. This energy difference accounts for the different spectral lines. Of interest is the fact that with these spin quantum numbers an electron would have to spin around two times in order to return to its 'starting point'.

Pauli concluded that each quantum state was only limited to one electron and that there are four quantum numbers, including spin up or down, for each. He developed his famous exclusion principle stating that when an electron within an atom had its four quantum numbers, no other electron could utilize that orbit. This meant that once a quantum state is occupied, another electron will only go to the next empty higher energy state---always from lower to higher. And thank goodness, for this gives each element a structure unique to itself. If this were not the case, electrons would always end up in the lowest energy state and solid matter as we know it would not exist.

Bohr and Pauli's work made possible a clear explanation of Mendeleev's periodic table of the elements. Bohr had two concerns when he developed his quantum theory: first, the explanation of Balmer's spectra, and second the clarification of the periodic table. Bohr pictured the electrons revolving in shells or orbits around the nucleus and suggested that the way the electrons configured in their shells explained an element's chemical and physical properties. Pauli clarified this issue with his famous exclusion principle. For example, there are some elements that are inert and are therefore incapable of combining with any other element. These elements, known as inert gases, include helium, neon, and argon. These elements have full electron shells.

Hydrogen has one electron. Lithium has three electrons, two filling up its first shell, so the third electron has to go to the next higher energy shell according to Pauli's exclusion principle. Sodium has eleven electrons and since the first shell has two and the second shells capacity is eight, this leaves the eleventh electron no place to go except to the next higher energy shell.

Those elements that have a single electron in its outer shell are readily capable of combining with another element of a similar configuration. This outer shell electron, involved in chemical changes, is called a valence electron.

On the other hand, the inert gases have their outer shell filled to capacity with the full complement of electrons and are not free to combine with other elements, hence the term inert.

Bohr laid the groundwork and Pauli's exclusion principle automatically made it possible to provide the complement of electrons for each shell. Two, eight, eighteen etcetera. Every electron knows where its neighbors belong and therefore cannot impinge on those spaces and must find its own place in the atom's architecture.

For such amazing insight that so revamped physical theory, Bohr is to be given credit for one of the great intellectual achievements of all time. But just as Newton's mechanics was revamped by Einstein, and Maxwell's electrodynamics was revamped by Planck and Bohr, Bohrs quantum theory would be replaced within the next twelve years.

Prior to Bohr's paradigm shifting insight, the physics world's underpinnings were the wave-particle controversy. This occupied the attention of physicists for generations and Bohr attempted to replace it by settling upon the particle or photon. But this was not enough to vanquish the wave completely and the proponents of the wave theory would not lie down and whimper, but rose to the attack, and the first one to do this was a French Prince.

PARTICLE AND WAVE

Louis De Broglie (1892-1987)
of France, was born into a family of great wealth. They were descended from royalty. Louis studied history, and his brother Maurice studied physics. Once, out of curiosity, Louis attended a physics conference. He became so fascinated that he abandoned history and embarked upon a serious, formal study of physics at the Sorbonne in Paris. The combination of a brilliant brother as a resource, his own personal intelligence, money, and time, worked its magic. He became a physicist who was unafraid to stick a needle in the dogma that light is a particle. Einstein, who stated that as far as light is concerned, one might have to consider a duality, was a strong influence on De Broglie who favored the explanation of light that has both particle and wave manifestations. De Broglie raised the question of the impossibility of an isolated quantity of energy not being associated with a certain frequency as in Einstein's $E=h\nu$. He suggested that the photoelectric effect, which knocked electrons out of orbit, and the double-slit experiment, which for light was compatible with waves, meant that propagation of a wave should be associated with a particle. In this case, we are talking about an electron particle, but De Broglie suggested it meant any particle such as an electron, photon, or proton or any other bit of matter, small or large, including a pitched baseball. However, a baseball would carry with it a wave of

such tiny wavelength (point thirty-three zero's 12) or 0.000000000000000000000000000000000012 that it is not measurable and thus cannot be confirmed experimentally. Therefore, it is only possible to perform wavelength experiments for particles of small mass like protons or electrons. De Broglie presented this theory as his Ph.D. thesis and proceeded to amaze his examiners, one of whom was the prominent physicist…

Paul Langevin (1872-1946)
who sent the thesis to Einstein who commented that De Broglie's work was exceptional and may well explain many of the mysteries of physics. After that recommendation, De Broglie's Ph.D. thesis was accepted. A blessing by Einstein opened many doors.

His work confounded many other physicists as well. Hadn't this issue been settled as far back as Thomson? An electron was a particle---period.

But a theory remains a theory until verified by experiment. Two physicists

Davisson (1881-1958) and
Germer (1896-1971)
both demonstrated an orderly interference pattern of the electron using the double slit experiment. They confirmed De Broglie's theory. Electrons are particles that obeyed the laws of Newtonian Mechanics, but since they were also waves, they would also have to obey the laws of waves as well.

Now the quantum theorists had to ponder the question as to whether the electron was, in essence, a wave moving around its orbit with its electric charge distributed about the circumference, or was a particle circling the nucleus and jumping from one orbit to the next.

In an atom, De Broglie's theory suggested that an electron wave of a distinct frequency moved around the nucleus. The wavelengths would have to fit precisely around the orbit in a discrete whole number as Bohr had suggested with his particle theory.

IT'S FUZZY DOWN THERE

Werner Heisenberg (1901-76)
was a German physicist, the son of a professor of Greek history. He was an excellent piano player lending credence to the theory that young musicians learn math better. His hobby was mountain climbing, and the location of his birth, Munich, gave him ample opportunity to exercise his talent.

He entered the University of Munich where he studied under Erwin Sommerfeld, and met a fellow student, Wolfgang Pauli, and started a close friendship that would advance the profession they both passionately loved.

In 1922, Heisenberg attended a lecture by Bohr in Munich, Germany. He raised an issue that so impressed Bohr, that Bohr requested that they take a long walk and discuss it. Heisenberg was greatly impressed by Bohr's humble demeanor and open-mindedness, and the fact that he welcomed criticisms of his theory. Bohr was taken with the genius of Heisenberg, and silently predicted that a new star was about to arrive on the physics scene. An indeed he did.

Heisenberg felt that the correspondence principle was an inherent weakness in the Bohr scheme, but in spite of that he still understood that Bohr's theory pushed physics in a direction of great importance for a complete understanding of the atom.

Heisenberg, also a philosopher, felt that it made no sense to talk about and try and prove concepts such as Bohr's orbits that could not be seen. The inner structure of an atom was invisible to the eye and any instrumentation of the day. The only thing known with a great degree of accuracy was the light, or spectral pattern associated with certain frequencies and amplitudes that emanated from the interior of the atom. He therefore believed that a truer picture would emerge if one tried to explain Bohr's approach through mathematics that would connect quantum numbers and the atom's energy state with what they knew about the frequencies and brightness of the spectrum.

He came up with a mathematical system that worked, but at the same time gave him pause. In essence, in order for his theory to work 1x2 would not equal 2x1. As we all know it does, but this little glitch was the only thing that troubled Heisenberg. Whereas a less determined scientist might have thrown up his hands in despair, Heisenberg plowed forward, and performed more mathematical manipulations that managed to keep his theory intact and consistent.

He showed his results to Pauli, who approved, and eventually to

Max Born, (1882-1970)
a famous physicist in his own right, and Heisenberg's professor at the University of Munich where Heisenberg became a privat-dozent (the lowest rung of the academic ladder).

Max Born recognized in Heisenberg's calculations an anomalous multiplication rule, a mathematical construct invented fifty years ago by an English mathematician named Cayley: matrix algebra, an array of numbers with rows and columns that obeyed certain mathematical laws. And included in these laws was the anomalous multiplication rule that fell into place in Heisenberg's calculations. Unknowingly Heisenberg had reinvented matrix algebra.

Pascual Jordan (1902-1980)
took Heisenberg's mathematics and transposed it into a complete matrix system. It worked perfectly. The frequencies and intensities of spectral lines were correct.

In contrast to Bohr's theory, however, there was nothing like orbits or rotating and spinning electrons to visualize. Rather Heisenberg's matrix system was purely a mathematical format albeit an excellent one. They named it matrix mechanics. Like Newtonian Mechanics developed by Newton, quantum mechanics was undergoing birth pangs, and its midwife was Werner Heisenberg.

The spectral patterns of other atoms could also be derived using Heisenberg's mechanics, but no one understood the small glitch of non-commutability (2x1 does not =1x2)...

Erwin Schrodinger (1887-1961)
was an Austrian physicist who believed that the De Broglie wave concept was closer to the mark.

Einstein's praise for De Broglie's work stimulated Schrodinger to action.

Schrodinger's did his best mathematical work in a private room in a pleasant, scenic resort in the Tyrolean Alps in the company of a girlfriend. Whoever the young lady was, history never gave her any credit for providing the environment that for Schrodinger was conducive to the furthering and refinement of quantum mechanics.

He attempted to develop a mathematical model that would allow an electron wave to fit perfectly and travel in one of Bohr's orbits. He claimed that electron waves occupied atomic orbits as opposed to the prevailing thought about particles in orbit around a central nucleus. These waves produced light by constructive interference as the peaks of wave crests merged together.

The smallest orbit would be one consisting of a single wavelength. He reasoned that an electron would not spiral into the nucleus because an electron could never fit into an orbit less than a single wavelength. It would also be possible for an electron to fit into other larger orbits with certain shapes, angles, tilts, and spins if the electron had a whole number of wavelengths allowing for a snug fit.

Schrodinger developed a mathematical formula able to solve problems by utilizing electron waves: wave mechanics. Taken together with Heisenberg's matrix mechanics, quantum mechanics was born. Schrodinger's wave function permeates the space around the atom's nucleus and it describes the behavior of the particles by a probability wave. This probability wave gives a high index of suspicion as to where the particle might be. He viewed the electrons in orbit in a kind of probability cloud, and not as discrete precise electron orbits of the planetary model.

Eventually Schrodinger showed that his theory and Heisenberg's theory would give the same results even though they were different mathematical constructs. There was a developing consistency to quantum mechanics.

As with all theories that resolve some problems, but leave some unanswered questions, Schrodinger could not explain the photoelectric effect or black body radiation on the basis of his continuous wave model.

Again, Max Born came to the rescue when he worked out mathematically that it was impossible to exactly clarify the various quantum states. All one could hope to do is develop a probability that a certain quantum state exists. Exact answers were not possible on the atomic level. This was better able to reconcile the duality concept of particles and waves and demonstrated that Schrodinger's equation could only describe the likelihood that an electron would be in a certain orbital position. Schrodinger's equation then, as effective as it was, did not have a distinct physical reality as did the electromagnetic wave of classical physics. Nevertheless, his equation has stood the test of time.

Paul Dirac (1902-1984)

was a British physicist who started his career as an electrical engineer. He changed to mathematics and physics, however, and became one of the premier physicists of all time. Many considered him the equal of Einstein.

His contribution was the development of a more general version of quantum mechanics that incorporated Heisenberg's matrix mechanics, and Schrodinger's wave mechanics. The three versions were equivalent. This all had the effect of placing quantum mechanics on a much stronger mathematical foundation. His successful theory expanded Schrodinger's work. Now fields as well as particles could both be quantized. Dirac's amazing work made him a physics figure to be reckoned with and he was now called forth to join the Copenhagen group of theorists.

Dirac's next contribution came in 1928 when he combined quantum mechanics with Einstein's special relativity. There was a flaw in Schrodinger's wave mechanics which could not take into account the fact that the electron, which moved so fast, could be better explained by Einstein's special relativity as opposed to Newtonian Mechanics. This posed great problems, but Dirac solved them by an amazing equation marrying quantum mechanics to relativity that also confirmed the previously developed concept of electron spin, and demonstrated that the electron could have positive energies as well as the well-known negative. In other words, his equation called for the prediction of a positive electron, or anti electron.

This concept was not considered seriously until...
Carl Anderson (1905-1991)
an American physicist, confirmed it. While measuring cosmic rays Anderson discovered the anti-electron (positron). Dirac demonstrated later that if an electron collided with a positron they would annihilate each other almost instantaneously releasing energy as gamma-rays according to Einstein's famous equation $E=MC^2$.

Dirac, in addition to placing quantum mechanics on a more solid footing, had also opened the door to a new era of anti-particle physics.

Clearly we cannot see electrons and we cannot see atoms. In order to make a determination of their characteristics we have to resort to such indirect methods as the use of a cloud chamber or examining spectral patterns. Whether we are dealing with waves or particles, they are invisible to the naked eye. Because of this, it is basically impossible to measure electrons accurately at the quantum level in the same manner that we can measure the mass of a billiard ball or the velocity of a bullet, or the precise location of a planet as it rotates around the sun. All these things are possible using Newtonian Mechanics at the macroscopic level, but at the sub-sub microscopic quantum level where a billion billion electrons do not weigh as much as a feather, such refinement breaks down. If one were to try to determine the exact position of a particle, any device used to make this determination would change the particle's velocity and therefore its momentum. Also, any device that might be used to determine a particles momentum would change the particles position. Therefore, the best that one could hope to achieve would be a very close approximation of the position and the momentum but never an exact determination. Such inaccuracies are the best hoped for at the quantum level.

Heisenberg developed this principle, which in part developed from his non- commutability rule and explained it mathematically by saying that the uncertainty of the position times the uncertainty of the momentum was never less than Planck's constant. In other words it is an impossibility to know the position or the momentum of a particle to a more exact degree then is spelled out by Planck's constant h.

In simpler language yet, what Heisenberg did was set out the physical limits to the observation of the electron. We can only be aware of the probability of the electron's position and its momentum

In spite of the fact that many scientists shunned the theory, it works. Time has not changed the concept. Even Einstein refused to accept the "uncertainty principle" which gave rise to his famous saying unchanged to the end of his life: "God does not play dice with the universe." Bohr finally told Einstein, "Please stop telling God what to do."

So, through all the brilliance of physicists, we are still finally left with a quantum mechanics that features probability and uncertainty as its main tenet. But, it has stood the test of time and provides experimentally confirmed observations. That's the final outcome and that's the way it is---until such time as something better comes along.

Jacob Bronowsky said: The progress of science is the discovery at each step of a new order that gives unity to what had long seemed unlikely. Faraday did this when he closed the link between electricity and magnetism. Clerk Maxwell did it when he linked both with light. Einstein linked time with space, mass with energy, and the path of light past the sun with the flight of a bullet; and spent his dying years in trying to add to these likenesses another, which would find a single imaginative order between the equations of Clerk Maxwell and his own geometry of gravitation.

Moving into the 1930's physics would take an unexpected turn. This would eventually usher in an era where humankind would call to question the future of life on earth.

THE NUCLEAR AGE: PROMISE OR DISASTER?

A number of experiments were performed that could have ushered in the nuclear era had the experimenters been better able to understand their results. In 1930 Bothe and Becker of Germany bombarded beryllium with alpha particles and interpreted the radiation liberated from the beryllium to be 'gamma rays'. Irene Joliet-Curie (1897-1956), and her husband Frederic Joliot-Curie (1900-1956) French physicists, propelled the 'gamma rays' liberated from beryllium into a sheet of paraffin. The paraffin, once struck by the 'gamma rays,' liberated protons. The Joliot-Curies assumed the 'gamma rays' to be extremely energetic in order to affect the much heavier proton.

James Chadwick (1891-1974)
was an English physicist who could not accept this explanation, and by a series of experiments proved that the 'gamma rays' were really the neutral particle: the neutron, often postulated in the past and now finally discovered. This breakthrough finally clarified the nucleus of atoms: they contained protons and neutrons.

Heisenberg immediately offered an interesting theory. He stated that the nucleus of atoms also contained electrons, and the neutron needed a proton and an electron whose positive and negative charge canceled each other out. In fact, he suggested that the nuclear force, required to hold positively charged protons together, resulted from the passage of an electron from a proton to a neutron. The passage back and forth of the electron would change the neutron to a proton and visa-versa. Heisenberg really postulated an exchange force due to an exchange of particles, and this concept would prove to be correct, but three years later,

Hideki Yukawa (1907-1981)
of Japan would replace the electrons with what he called mesons. Physicists were clarifying the powerful nuclear force.

There was one question requiring an answer, however. Since the nuclear force was so powerful, how did alpha particles emitted by heavy elements like uranium manage to break through this powerful nuclear force and liberate itself through the well-known spontaneous process known as alpha decay?

George Gamow (1904-1958)
a Russian physicist, showed how this mechanism worked. Alpha particles were able to "quantum tunnel" out by borrowing some energy and giving it back in an infinitesimally short time-frame consistent with Heisenberg's uncertainty principle.

Generations of physicists established the groundwork for men like the Italian physicist,

Enrico Fermi (1901-1954)
who unleashed the power of the atom, or to put it more exactly, the power residing in the nucleus of the atom. The potential of the newly discovered neutron became clear, and Fermi started to bombard all known elements with this new magic bullet. He proved that if the neutrons slowed by passing them through paraffin, they became more effective in penetrating the nucleus, and the result was a new isotope of the element that would prove to be unstable and decay rapidly. Fermi bombarded uranium with neutrons and developed new isotopes called transuranics, which he felt were higher than uranium on the periodic table---or so he thought.

As Fermi was doing his research, Adolf Hitler had suggested that Mussolini reign in his Jews. Fermi was not Jewish, but his wife and children were, so he and his family fled Italy and immigrated to the United States.

Fermi made one error in not analyzing the transuranics he made by bombarding uranium with neutrons, but

Otto Hahn (1879-1968)
of Germany, did not repeat the mistake. He and...
Fritz Strassman (1902-1980)
discovered that one of the products of neutron bombardment of uranium was barium. The problem was that barium was one-half the atomic weight of uranium. As a radiochemist, Hahn had understood

that such a result was impossible, so he consulted with his colleague...

Lisa Meitner (1879-1968)
who was a physicist of Jewish background who had fled the Nazis. She, and her nephew,
Otto Frisch (1901-1979)
confirmed that this was indeed the case: uranium split in half. In other words, using a word coined by Frisch—a word he borrowed from biology—the uranium atom fissioned. When a uranium nucleus absorbed a neutron it becomes unstable and splits in half. In so doing, it liberates an enormous amount of energy.

When word got out of this momentous discovery, physicists of the world were galvanized into action. Fermi went back to work.

Leo Szilard (1898-1964)
a Hungarian Jewish physicist who also was forced to emigrate to the United States, knew that if uranium fissioned by being bombarded by a proton, the fissioning process would release two more protons. These two protons then would fission two more nuclei with the liberation of four more protons, etcetera. Within milliseconds there would be about eighty splits. In a controlled process enough energy could be produced to power great cities. In an uncontrolled process a 'nuclear bomb' could be produced with an explosive power such as the world had never seen. The race was on. Hitler had to be beaten. The world's future was at stake.

Enrico Fermi, working with many home grown and émigré physicists at the University of Chicago, then proved a controlled chain reaction was possible. Leo Szilard convinced Albert Einstein to write President Roosevelt a letter urging him to embark upon science's greatest quest. The Manhattan project was born, and under the direction of...

J. Robert Oppenheimer (1904-1967)
an American physicist, and
General Groves (1896-1970)
of the United States Army, the nuclear bomb was born. The race for mastery of the world had begun. Civilization was on the threshold of good and evil.

END GAME

So what have we learned so far?

The simplest element is hydrogen:

A hydrogen atom is made up of a proton and an electron. There are other forms of hydrogen with different amounts of neutrons called isotopes (the same is true for most elements).

The most common form of hydrogen has one proton in the nucleus and one electron circling the nucleus in the distant periphery. A proton is a particle with a positive charge of electricity and an electron has a negative charge of electricity. When Theophrastus wondered if the attractive force generated by rubbing amber and other substances was a universal force helping to explain all matter, he was right.

An atom of hydrogen consists of an electron revolving around a proton just as our earth revolves around the sun. And if we would enlarge the atom so that the proton was the size of the earth, then the electron would be revolving around the proton from a distance almost twice as far as the earth is from the sun. As was mentioned before, the electron weighs about 1/1900th that of the proton and is smaller in size by about half.

Clearly, we've just heard a description of an atom that is made up of two components separated by a vast distance. Does that mean that an atom is mostly empty space? The answer, which is yes, makes one wonder how anything can be solid which is mostly empty space, for isn't much of our world, all of which is made up of atoms, solid? And if human beings are also made up of atoms doesn't that mean that we are mostly empty space? The answer is again yes, and if we got rid of all that empty space we would suddenly become the size of the question mark at the end of the preceding sentence. So how does all solid matter in the universe stay solid? The answer is that the electrons revolve around the nucleus with incredible speed forming a blur or shield. Also we have learned that this speed is so fast that if we stood outside of the electron (or in many elements---electrons) and tried to get at the nucleus, we would be sliced up in little pieces by the whizzing electron particle or particles (or as we learned from De Broglie, a well- fitting orbital electron wave).

We started with the simplest atom---hydrogen, and hinted that other atoms had different structures. That's true as long as you realize that the basic structure remains a nucleus with a revolving electron as in hydrogen or revolving electrons as in all other elements.

In order to discuss the make-up of other elements we have to remember that James Chadwick, a British physicist, confirmed the fact that there was another particle within the nucleus of an atom, and that is the neutron, so named because of its electric neutrality. Except for the predominant form of hydrogen, the nucleus of all other atoms consists of protons and neutrons. This now gives us three particles (or in the case of an electron---a particle and/or wave) making up all the atoms of the universe; at least this universe as we know it in the 1930's.

A neutron is a particle that has almost the same weight and size as the proton, but since it is electrically neutral, it isn't affected by the electrical charge of its nuclear neighbor, the proton.

Let's answer two questions before we go on. Since opposite charges attract, why doesn't the electrically negative electron speed right into the positively charged nucleus? In way of review, the answer was, according to Bohr, an electron does not give up energy when rotating in a stable orbit, or according to DeBroglie, an electron wave fits exactly within the orbit and with this snug fit gives off no energy.

Ok then, if that explains the electron, how does one explain the fact that in elements that have multiple protons, why do these protons not repel each other as like charges are supposed to do? The answer brings to the fore another important concept which has great relevance for our story. The protons are bound together by a nuclear force (see mesons above). As you can imagine, this force is very powerful and important, for without it atoms could not exist and we would all be part of some primordial soup of energy waiting to evolve.

This force, called the strong nuclear force, acts only within the nucleus of an atom at a range of only 0.000000000000001 meters (one quadrillionth of a meter), and acts as a glue holding the nucleus in one piece. At the tiny range within which it acts, it has the ability to push protons apart if they come too close, or push them together if they get too far apart. To put it in context, this strong nuclear force is 100,000,000,000,000,000,000,000,000,000,000,000 (one hundred unodecillion, or 100 billion billion billion billion times greater than the force of gravity. Inside the nucleus the strong nuclear force is also one hundred times stronger than the electromagnetic force, and one million times stronger than the weak force which is a force responsible for radioactive decay.

In order for the strong nuclear force to work, the protons and neutrons within the nucleus must be very close together, actually no farther than the diameter of one of the particles. When they are this close there is an exchange of a particle called a meson which bounces back and forth between the protons like a ping pong ball, and this bouncing back and forth keeps the nucleons together. If they can't get this close, the strong force doesn't have the power to keep the protons together and they will move apart as a result of the electromagnetic force of repulsion.

It is the neutrons in the nucleus which help to reduce the force of repulsion between protons. Since neutrons have no charge they don't add to any repulsion in the nucleus and they help to keep the protons apart. This enables the protons to be shielded somewhat from the repulsive force of other protons. Since the neutrons also participate in the meson exchange, they too are the source of some of the strong force.

With this knowledge, one can understand why it is easier to bombard a nucleus with a neutron then a proton. A neutron will not be repulsed (it has no charge) as it speeds toward a nucleus and therefore can more readily break the electrostatic repulsion barrier and become incorporated into the nucleus, or otherwise do their nuclear mischief so critical for the nuclear era.

The protons and neutrons within the nucleus are lumped together under the title of nucleons. (If we jump forward forty years we would learn that neutrons and protons are made of quarks---more later.

We'll now discuss the composition of other atoms in the periodic table.

After hydrogen, the next element is helium. Its most abundant form has two protons and two neutrons and two negatively charged electrons in orbit.

The third element is lithium and its most common form has three protons, three neutrons, and three electrons.

The fourth element is beryllium. Its most common form has four protons, five neutrons, and four electrons.

The fifth element is boron. It has five protons, six neutrons, and five electrons.

The sixth element is carbon. It has six protons, six neutrons, and six electrons.

As we go up the periodic table we note that the number of neutrons becomes greater, so that by the time we get to uranium we have an atom consisting of ninety-two protons, and 146 neutrons. These extra neutrons have relevance, as we shall see later.

If an atom loses an electron, (and this can happen from outer orbits) the atom will now have a positive charge one greater than the negative charge. If an atom gains an electron, the negative charge will be one greater than the positive charge. Such atoms, which are no longer electrically neutral, are called ions and the process of loss or gain of an electron is called ionization.

We've touched upon the strong force that keeps the nucleus together. Clearly, such a powerful force means that there is an incredible amount of energy tied up in the nucleus of an atom. To better understand this energy let us assume that two protons have been increased in size to that of two single jelly beans. Now let's decide to have one attract the other by bringing them closer together. If we did that, and tried to measure the force and speed of the attraction, it would be the equivalent of two freight trains, each weighing two thousand tons or four million pounds hurling together at close to the speed of light. Such is the nature of the force tied up in the nucleus of the atom, and this was quantified by Albert Einstein when he developed his famous equation $E=MC^2$.

To attempt to come to grips with this concept let us go back in time about fifteen billion years to the Big Bang. If we accept the Big Bang concept as explaining the origin of the universe, then we accept the fact that there was an enormous amount of energy virtually concentrated at a point. This energy liberated itself by an explosion that traveled through a developing time and space. As the milliseconds and eons passed, this energy coalesced into the matter of our universe: planets, stars, and galaxies. So is it a surprise that mass is in reality a coalesced form of energy? Mass and energy are two different forms of the same thing. Since some of the energy of the primordial universe has become mass, why could not the mass turn back into energy? Einstein showed us that if we destroy mass we will get back an amount of energy equivalent to the mass lost times the square of the speed of light.

To turn this into meaningful numbers, let us assume that we start with a mass of one gram. A gram is a little more than one thirtieth of an ounce. You have learned that the speed of light is one hundred eighty-six thousand miles per second. Since we are working with a gram of mass, let's turn the one hundred eight-six thousand miles per second into centimeters per second so as to allow us to stay with the metric system. One hundred eighty-six thousand miles per second works out to thirty billion centimeters per second. Squaring this we get nine hundred quintillion centimeters per second or nine hundred million million million centimeters per second or nine hundred billion billion centimeters per second depending upon what way of saying it is more meaningful to you.

Now, using Einstein's formula, we get $E =$ one gram times nine hundred million million million centimeters per second which equals nine hundred million million million ergs. What is an erg?

Consider it a measure of energy and suffice it to say that this many ergs is plenty of energy. In fact, it's enough that if you utilized this amount of energy you would win a tug of war contest against sixty five billion horses. Another way of trying to understand it is to say that the one gram of mass would liberate an amount of energy equivalent to burning fifty thousand tons or one hundred billion pounds of coal. Pretty incredible!

Another important point to comprehend is that the strong nuclear force, which as we discussed can only make itself felt over tiny distances (one ten million millionth of a centimeter), is in contrast to the electromagnetic force which decreases slowly over distance according to the inverse square law. This difference is responsible for the occurrence of fission taking place in large nuclei such as uranium. Fission is the splitting of a heavy nucleus into two or more pieces as a result of the impact of a neutron on a nucleus. Plutonium, uranium, and thorium are the elements capable of splitting in this manner. If this sounds like what you have heard in reference to nuclear bombs or nuclear energy you are correct.

It's important to understand that in a large nucleus, due to the nature of the strong nuclear force and the fact that each proton or neutron only has a small finite number of such fixed neighbors, a proton or a neutron only feels the force of its close neighbors. Because of the long range of the electromagnetic force, all protons in the nucleus are affected by each other. Therefore, the more protons there are the more the repelling electromagnetic force becomes significant. The result of this is that such large nuclei are inherently unstable and therefore fissionable. This is a critical concept to understand as we learn about the effort to build the nuclear bomb.

The last topic for discussion is an isotope. Let us take the hydrogen atom as the first example. It has been determined that the hydrogen atom not only consists of one electron revolving around one proton, but there are also two other hydrogen atoms. One known as deuterium has one electron revolving around a nucleus consisting of one proton and one neutron. The second isotope of hydrogen has one electron revolving around a nucleus consisting of one proton and two neutrons: Tritium. There's very little deuterium and tritium, but they are there. The atoms are all electrically neutral but have different weights because of the extra neutrons. Other elements have isotopes as well, but we'll jump to uranium because of its relevance to our story. Uranium is labeled ^{238}U because of the number of nucleons in the nucleus. But there is also ^{235}U, ^{233}U, and ^{234}U. By definition then, the ^{235}U has three less neutrons then ^{238}U. ^{235}U is critically important, as it is the isotope of uranium that is most radioactive and is most relevant to our story.

As atoms break down it becomes clear that the original atom will become less and less over time. How much of the original substance will be left over a given period of time? The way this is calculated is by the concept of half-life. In other words if it takes one year for half of the original atoms to break down leaving half of the original material, then in the next year half of the half that remained will break down. This will leave one-quarter of the original material, which will break down in the next year leaving one-eighth of the original material, and so ad infinitum.

It turns out that the half-life of ^{238}U is 4.5 billion years, and the half-life of its 235 isotope is seven hundred million years. The half-life of Thorium 232 is 14 billion years. The half-life of other substances can be only fractions of a second to any amount between these extremes. The 235 isotope of Uranium is the 'enriched' uranium you read about so critical to the development of the nuclear bomb. Today it represents only 0.7 percent of the total uranium. If we go back several billion years it was three percent, but because of its shorter half-life it is much less today.

This concludes the review of physics from its origins over two thousand years ago to the onset of the atomic era during World War II. We hope you enjoyed the trip and the famous scientists you met along the way.

Now we will embark upon the trends in Physics from World War II to the modern era.

WHAT HAPPENED TO AIR, EARTH, FIRE AND WATER?

HOW THE UNIVERSE WORKS
SUBATOMIC PARTICLES

As previously described, the first particle discovered was the electron, identified by
J.J. Thomson in 1877 and
Ernest Rutherford, in 1911.
They recognized that the nucleus of hydrogen was occupied by a single proton surrounded by a rapidly revolving electron, just as the earth rotates around the sun. This configuration was the opening salvo in what would be an ever evolving complexity of atomic configuration.

By 1928 the relativistic quantum theory of P.A.M. Dirac made its presence felt. Quantum theory is a subset of physics that deals with the physical behaviors at the molecular, atomic and sub-atomic levels of the physical world. Dirac hypothesized the existence of a positively charged electron. Four years later in 1932, this particle, named the positron, was discovered. However, other elementary particles not found in ordinary atoms began to appear in a nucleus with too many protons or neutrons. Neutrons, discovered in 1932, have no electric charge and a mass slightly larger than that of a proton. Neutrons and protons constitute the nucleus of all atoms above the weight of a hydrogen atom.

In the nuclear world of larger atomic structure, beta decay can occur in a nucleus with too many protons or neurons. During this process of beta decay, the proton is converted into a neutron and visa-versa. A beta particle is an electron or positron (electron with positive electric charge, or antielectron). Beta decay happens when a nucleus laden with too many protons or neutrons reacts, and one of the protons or neutrons is transformed into the other as above described.

From this beginning to this current year of 2017, physicists have determined that our universe is made up of 24 elementary particles, all interacting with each other via four fundamental forces transmitted by quantum carriers, also elementary particles. These particles discovered one by one in a period of approximately 125 years led to an understanding of physics at the atomic and sub-atomic level, rewarding many with Nobel prizes. Now that the 24 particles are all in place, through their complicated interactions we have a functioning world within which we can live and interact.

There are four forces on our world: gravity, electromagnetic, strong, and weak.

Gravity

Range of action is infinite. It encompasses the universe. It acts throughout outer space with a strength characterized as 10 to the minus 38th power (compared to a strong force of one in the atomic nucleus. Gravity acts on everything including energy. Its carrier particle, known as the graviton, has never been observed.

Strong

This is the force that holds the nucleus of atoms together. It must be super strong to accomplish this task as it acts at a range of ten to the minus 14 meters with a strength labeled as one and affecting only quarks and gluons. The carrier particle is the gluon.

Electromagnetic

This force is an interaction between electrically charged particles such as in electric fields, magnetic fields and light. Its range is infinite, its strength is 10 to the minus 2 power, and it affects charged particles and the photon, which is the carrier particle.

Weak

The weak force is responsible for some forms of radioactivity, causes decay of unstable subatomic particles, and initiates the sun's nuclear fusion reaction. Particles interact via the weak force by exchanging force-carrier particles known as W and Z, heavy particles with masses of 100 times the mass of a proton. This heaviness is responsible for the short-range of the weak force. Its effectiveness is short and extremely powerful; only ten to the minus 17 meters.

What had developed over time was a classification of the make-up of our world at the atomic and subatomic level and how it all comes together; complicated and difficult to follow, but an amazing intellectual development of humankind, the only creature of the thousands of life forms on this planet with the ability to contemplate such issues. I tried to summarize its intricacies in the brief comments above, so I hope the reader will just appreciate the great intellectual accomplishments of the scientists who could indeed put it all together in a more than century long step-by-step discovery process. Our role is to sit back and be amazed!

HISTORY OF THE UNIVERSE

Well beyond 2,637 years ago, primitive man looked up at the sky and did not wonder about this large changing shaped thing which rose over the horizon every evening and added some ability to see in the dark. There were other more important things to contemplate, such as where the next meal was coming from. But God in his wisdom knew there was promise in this creature (amongst all the others on earth), and decided to give them a slowly increasing ability to think. Eventually that choice, with God's directions, blossomed out into the thinking human being God had envisioned who would soon be able to contemplate the world around him and wonder as to all those sparkling lights in the evening sky. What is up there? How did it get there?

Well into the future when they were ready, God gave this wisdom to a Belgian priest (it figures) named

Georges Lemaitre (1894-1966) the first to suggest that the universe began from a single primordial atom that erupted into our universe and sped away in all directions. Lemaitre labeled this concept the Big Bang. (How did he figure this? Did God whisper in his ear?).

It did not take long before

Edwin Hubble (1889-1953)
proved that galaxies in outer space were speeding away from each other lending credence to Lemaitre's Big Bang theory. In addition, the discovery of cosmic microwave radiation by

Arno Penzias 1933 and
Robert Wilson (1927-2002
is thought to be a remnant of the Big Bang, again adding credence to the Big Bang theory. This radiation is the oldest radiation known and it holds the secret of the universes earliest moments.

God was not yet ready to give up all his secrets, such as how did the big bang happen many billions of years ago? And how did this dense mass smaller than a proton explode in the first fraction of a second, and expand away from its center point at an amazing speed (faster than light?)...so that within .00000000000000000000000000000000001 second... it had become the size of a grapefruit.

As put by one of the deep thinkers of this era..."the universe came from almost nowhere in no time."

It was thought that the modern day universe, as it got further away from its Big Bang point of origin, was slowing down. No doubt because scientists felt that gravity's force would slowly pull the expanding universe back together eventually, but now, a great discovery ensued!

By virtue of advanced telescopic wizardry in the hands of God's chosen humans who looked far out toward the slowing down distant galaxies, these same humans now found that the galaxies were not slowing down, but rather were accelerating apart from each other at an increasing speed! What made the universe speed up!?

Everyone agrees that our universe is currently 15 billion years old. Why did it change direction and the galaxies move away from each other faster instead of slowing down? This phenomenon is attributed to dark matter, a form of matter which permeates all space and contributes 68.3 percent of the total energy in the present day universe and overcomes gravity and accelerates the universes expansion. Does this mean that the universe will now expand forever and eventually "burn out" and form what could be a vacuum?

Don't hold your breath, however, because we are being told by the deep thinkers that that will happen in about ten thousand trillion trillion trillion trillion trillion trillion trillion trillion years. That is a 1 followed by 100 zeros.

The great minds are wrestling with this phenomenon, and I am sure it will continue to flourish in human thought processes as long as humankind remains God's chosen vehicle to contemplate such issues.

Isn't it interesting that the two greatest advances in Physics since the end of World War 2 (1945) represent advances involving the very smallest (atomic physics) and the very largest (the universe)!

Stay tuned.

BIBLIOGRAPHY

Asimov, Isaac, Atom. Journey Across The Subatomic Cosmos, Plume, New York, 1992.

Asimov, Isaac, Understanding Physics: The Electron , Proton, and Neutron, Signet, New York, 1966.

Bentinck, Henry, Anyone Can Understand the Atom, Max Parrish, London, 1965.

Bernstien, Jeremy, Hitler's Uranium Club, Copernicus Books, New York, 2001.

Bodanis, David, E=MC2, Berkley Publishing Group, New York, 2000.

Dogigli, Johannes, The Magic Of Rays, Alfred A. Knopf, New York, 1961.

Engelmann, Bernt, Germany Without Jews, Bantum Books, New York, 1979.

Feinberg, Gerald, What Is The World Made Of? Anchor press, Garden City, New York, 1977.

Fermi, Laura, Illustrious Immigrants, The University of Chicago Press, Chicago, 1968.

Frayn, Michael, Copenhagen, Anchor Books, New York, 1998.

Freudmann, Lillian C. Antisemitism in the New Testament, University Press of America, New York, 1994.

Gamow, George, Thirty Years That Shook Physics, Doubleday & Company, Inc., Garden City, New York, 1966.

Hall, Walter Phelps, and Davis, William Stearns, The Course Of Europe Since Waterloo, Appleton Century-Crofts, Inc., New York, 1951.

Heisenberg, Werner, Physics and Philosophy, Prometheus Books, Amherst, New York, 1999.

Hoffman, Banesh, The Strange Story Of the Quantum, Dover Publications, Inc., New York, 1947.

Kurtzman, Dan, Day Of The Bomb Countdown to Hiroshima, McGraw-Hill Book Company, New York, 1986.

McEvoy, J.P., Zarate, Oscar, Quantum Theory. Totem Books, USA, 1997.

May, Arthur James, Europe And Two World Wars, Charles Scribner's Sons, New York, 1947.

Polkinghorne, J.C., The Particle Play, W.H. Freeman and Company, Oxford and San Fransisco, 1979.

Powers, Thomas, Heisenberg's War. The Secret History Of The German Bomb, Da Capo Press, 1993.

Rauschning, Hermann, The Voice Of Destruction, G.P. Putnam's Sons, New York, 1940.

Reeves, Pamela, Ellis Island, Barnes and Noble Books, New York, 1998.

Rhodes, Richard, The Making Of The Atomic Bomb, Simon and Schuster, New York, 1986.

Shirer, William L., The Rise And Fall Of The Third Reich, Simon & Shuster, New York, 1959.

Segre, Amelio, From X-Rays To Quarks, W.H. Freeman And Company, San Francisco, 1976.

Transnational College of Lex, What is Quantum Mechanics? A Physics Adventure, Language Research Foundation, Boston, 1996.